ANATOMY, PHYSIOLOGY & DISEASE
FOR THE HEALTH PROFESSIONS

Third Edition

STUDENT WORKBOOK FOR USE WITH

Kathryn A. Booth, RN-BSN, RMA (AMT), CPT, CPhT, MS
Military to Medicine, INOVA Health System
Falls Church, VA

J. Virgil Stoia, DC
College of DuPage
Glen Ellyn, IL

Student Workbook for use with
Anatomy, Physiology, and Disease for the Health Professions, Third Edition
Kathryn A. Booth and J. Virgil Stoia

Published by McGraw-Hill, a business unit of The McGraw-Hill Companies, Inc., 1221 Avenue of the Americas, New York, NY 10020. Copyright © 2013 by The McGraw-Hill Companies, Inc.

All rights reserved. Printed in the United States of America. Previous editions © 2008 and 2009.

The contents, or parts thereof, may be reproduced in print form solely for classroom use with Anatomy, Physiology, and Disease for the Health Professions, Third Edition, provided such reproductions bear copyright notice, but may not be reproduced in any other form or for any other purpose without the prior written consent of The McGraw-Hill Companies, Inc., including, but not limited to, in any network or other electronic storage or transmission, or broadcast for distance learning.

1 2 3 4 5 6 7 8 9 0 QDB/QDB 1 0 9 8 7 6 5 4 3 2
ISBN 978-0-07-747517-8
MHID 0-07-747517-8

Cover credit: *Illustration Copyright© 2011 Nucleus Medical Media, All rights reserved.*

www.mhhe.com

contents

Unit 1: The Human Body and Disease 1

1 Concepts of the Human Body *1*

2 Concepts of Chemistry *9*

3 Concepts of Cells and Tissues *17*

4 Concepts of Disease *27*

5 Concepts of Microbiology *35*

6 Concepts of Fluid, Electrolyte, and Acid–Base Balance *45*

Unit 2: Concepts of Common Illness by System 51

7 The Integumentary System *51*

8 The Skeletal System *61*

9 The Muscular System *73*

10 Blood and Circulation *83*

11 The Cardiovascular System *93*

12 The Lymphatic and Immune Systems *103*

13 The Respiratory System *111*

14 The Nervous System *121*

15 The Urinary System *131*

16 The Male Reproductive System *141*

17 The Female Reproductive System *149*

18 Human Development and Genetics *157*

19 The Digestive System *167*

20 Metabolic Function and Nutrition *179*

21 The Endocrine System *187*

22 The Special Senses *199*

iii

To the Student

If you have chosen a health career for your life's vocation, you will need a solid foundation in anatomy, physiology, and disease. Virtually all courses in the health care and clinical sciences start with fundamental knowledge of how the human body is structured and how it works, as well as common diseases and disorders of the body systems. This workbook is designed to reinforce the concepts presented in the textbook. By diligently working through the many questions, labeling exercises, case studies, and pathophysiology sections, you will become more proficient and comfortable with the knowledge that you need to enter your chosen profession in health care.

Learning Outcomes. The Learning Outcomes presented in the text were the starting point for the questions and exercises presented here. They are the basis for the sections presented in the workbook.

Vocabulary Review. This section tests your knowledge of important terms introduced in the chapter.
- *Matching* exercises at the beginning of each workbook chapter will reinforce the key and essential terms you have learned.

vocabulary review

MATCHING

Match the key terms in the right column with the definitions in the left column by writing the letter of the correct answer in the space provided.

_____ 1. Anything that takes up space
_____ 2. Neutrally charged particles found in the center of an atom
_____ 3. Negatively charged particles surrounding the nucleus
_____ 4. The number of protons in an element
_____ 5. Contains the genetic information of the cell
_____ 6. Combination of two or more different atoms of different elements
_____ 7. Unit of matter that makes up a chemical element
_____ 8. A branch of chemistry dealing with the chemistry of life
_____ 9. What holds atoms together
_____ 10. Small molecules combine to form larger molecules
_____ 11. The breakdown of complex molecules into simpler molecules with the release of energy
_____ 12. Atoms with the same atomic number, but different atomic weights
_____ 13. An atom or group of atoms with a positive charge
_____ 14. Used to store energy for cells
_____ 15. The center of the atom where protons and neutrons are located

a. Anabolism
b. Anion
c. Atom
d. Atomic number
e. Atomic weight
f. Biochemistry
g. Catabolism
h. Cation
i. Chemical bond
j. Compound
k. DNA
l. Electrons
m. Isotopes
n. Matter
o. Neutrons
p. Nucleus
q. Protons
r. Ribosomes
s. RNA
t. Triglycerides

Content Review. This section tests your knowledge of learning outcomes and concepts introduced in the chapter. Formats for these exercises include the following:

- *Multiple Choice.* The multiple choice section gives you the opportunity to choose the correct answer while understanding why the other choices are not correct.

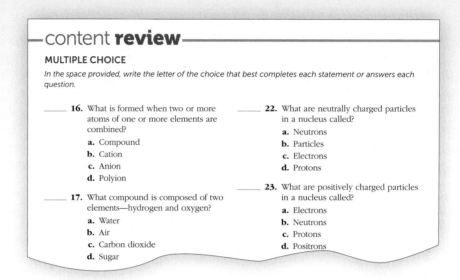

- *Fill in the Blanks.* The exercises will challenge you to recall the key information that you will want to master.
- *Short Answer.* The short answer section in each chapter of the workbook will help you become comfortable writing the ideas that have been presented in this course.
- *Labeling.* Labeling of diagrams is a visual reinforcement of the knowledge that you are acquiring and helps you begin to apply the content.

LABELING
Follow the directions and write the answers on the lines provided.

51. Identify these terms in the following figure by writing them on the lines provided: *covalent bond, electron, neutron.*

a. _____

b. _____

c. _____

vi   To the Student

Critical Thinking/Application. The critical thinking/application section in the workbook will do just that. It will cause you to think in an analytical manner and apply essential concepts for a solid understanding of the anatomy and physiology of the human body.

Case Studies. The case studies are a fun way of testing yourself by considering real-life situations that you may be confronted with in the health care setting.

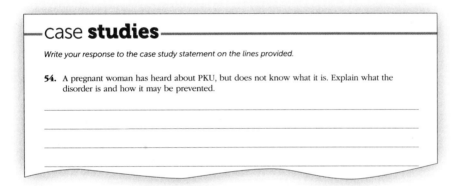

Pathophysiology. The pathophysiology section tops off each workbook chapter by putting it all together. It emphasizes the need for a thorough understanding of anatomy and physiology in order to better understand human disease. You will then be better able to understand and help individuals that will be placed in your care.

UNIT 1 The Human Body and Disease

Concepts of the Human Body

1

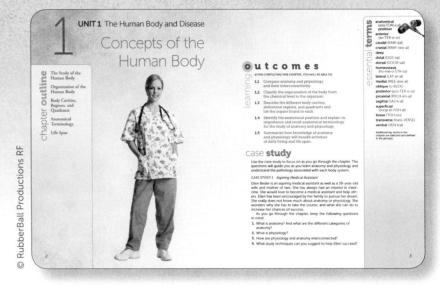

vocabulary review

MATCHING

Match the key terms in the right column with the definitions in the left column by writing the letter of the correct answer in the space provided.

_____ 1. The relative consistency of the body's internal environment

_____ 2. A group of similar cells that together perform a specific function

_____ 3. The body standing upright and facing forward with the palms of the hands facing forward

_____ 4. A plane that divides the body into left and right portions

_____ 5. Approaching or close to the head

_____ 6. The front of the body

_____ 7. The back of the body

_____ 8. Closer to the surface of the body

_____ 9. Pertaining to the thigh

_____ 10. Being closer to the trunk or a specified part

_____ 11. Away from the midline of the body

_____ 12. Closer to the midline

_____ 13. Away from the head

_____ 14. A plane that divides the body into superior and inferior portions

_____ 15. Away from the surface of the body

a. Anatomical position
b. Caudal
c. Cranial
d. Deep
e. Distal
f. Dorsal
g. Femoral
h. Homeostasis
i. Lateral
j. Medial
k. Proximal
l. Sagittal
m. Superficial
n. Tissue
o. Transverse
p. Ventral

1

content review

MULTIPLE CHOICE

In the space provided, write the letter of the choice that best completes each statement or answers each question.

_____ 16. The _____ cavity contains the heart, lungs, aorta, esophagus, and trachea.
   a. Dorsal
   b. Abdominal
   c. Thoracic
   d. Pelvic

_____ 17. A _____ plane divides the body into inferior and superior portions.
   a. Transverse
   b. Frontal
   c. Sagittal
   d. Midsagittal

_____ 18. The ankle is _____ to the foot.
   a. Distal
   b. Inferior
   c. Proximal
   d. Lateral

_____ 19. Which directional term is used to describe front to back?
   a. Mediolateral
   b. Anteroposterior
   c. Posteroanterior
   d. Anterolateral

_____ 20. The _____ body cavity contains the brain and spinal cord.
   a. Dorsal
   b. Ventral
   c. Thoracic
   d. Pelvic

_____ 21. In anatomical position, the body is
   a. Erect, with the head, feet, and palms all facing anteriorly
   b. Erect, with the head, feet, and palms all facing posteriorly
   c. Prone, with the head, feet, and palms all facing anteriorly
   d. Supine, with the head, feet, and palms all facing anteriorly

_____ 22. The shoulder is _____ to the elbow.
   a. Proximal
   b. Distal
   c. Superficial
   d. Intermediate

_____ 23. The abdominal and _____ are separated by the diaphragm.
   a. Pelvic cavities
   b. Thoracic cavities
   c. Dorsal cavities
   d. Ventral cavities

_____ 24. The appendix is found in the
   a. RUQ
   b. LUQ
   c. RLQ
   d. LLQ

_____ 25. Of the following, which is the simplest level of organization?
   a. Tissue
   b. Organism
   c. Organ
   d. Organ system

_____ 26. Which one of the following organ systems is involved with locomotion of the organism?
   a. Muscular system
   b. Digestive system
   c. Reproductive system
   d. Integumentary system

_____ **27.** Which one of the following is *not* a basic tissue type?
   a. Bone
   b. Muscle
   c. Epithelium
   d. Nervous tissue

_____ **28.** The terms *squamous, cuboidal,* and *transitional* apply to
   a. Epithelium
   b. Nervous tissue
   c. Skeletal muscle
   d. Cardiac muscle

_____ **29.** The bladder and internal reproductive organs are located in the
   a. Pelvic cavity
   b. Dorsal cavity
   c. Cranial cavity
   d. Abdominal cavity

_____ **30.** The spleen is part of the
   a. Lymphatic system
   b. Respiratory system
   c. Digestive system
   d. Integumentary system

FILL IN THE BLANKS

In the space provided, write the word or phrase that best completes each sentence. Not all words or phrases are used.

31. _____ is defined as the relative consistency of the body's internal environment.

32. The _____ system serves as a sense organ for the body and consists of the skin, hair, nails, and sweat glands.

33. The _____ system provides protection and support and allows body movements.

34. The _____ system helps maintain posture and produces body heat.

35. The tonsils belong to the _____ system.

36. The pharynx is part of the digestive and _____ systems.

37. In the male, the urethra is part of the urinary and _____ systems.

38. The spinal cord is part of the _____ system.

39. _____ describes the study of the body structures.

40. The study of each structure's function is _____.

a. Anatomy
b. Cardiovascular
c. Endocrine
d. Homeostasis
e. Integumentary
f. Lymphatic
g. Muscular
h. Nervous
i. Physiology
j. Reproductive
k. Respiratory
l. Sensory
m. Skeletal
n. Vascular

CHAPTER 1 Concepts of the Human Body

SHORT ANSWER

Write the answer to each statement on the lines provided.

41. For each body cavity, give its major divisions.

    a. Dorsal cavity _____

_____

    b. Ventral cavity _____

_____

    c. Abdominopelvic cavity _____

_____

42. Describe the organization of the body from the chemical level to the organism.

_____

_____

_____

_____

LABELING

Follow the directions and write the answers on the lines provided.

43. Using the following figure, write the terms that describe the relationship of one point of the body to another point on the lines provided.

    a. Point (a) in relation to point (c) _____

    b. Point (e) in relation to point (g) _____

    c. Point (l) in relation to point (k) _____

    d. Point (b) in relation to point (a) _____

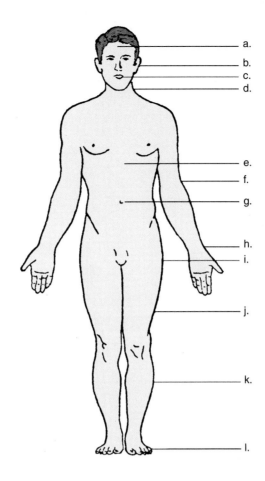

44. Using the previous figure, write the correct anatomical term for each labeled body area on the lines provided.

a. _____

b. _____

c. _____

d. _____

e. _____

f. _____

g. _____

h. _____

i. _____

j. _____

k. _____

l. _____

45. The following illustration shows the planes into which the body can be divided. Write the correct spatial terms on the lines provided.

a. _____

b. _____

c. _____

d. _____

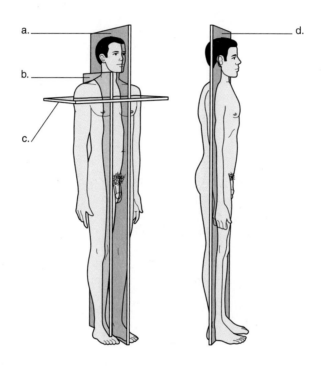

critical thinking / application

CRITICAL THINKING

Write the answer to each question on the lines provided.

46. Why are anatomy and physiology commonly studied together?

_____

_____

_____

_____

6 UNIT 1 The Human Body and Disease

47. Why is it important for a health care professional to have a basic understanding of anatomy and physiology?

APPLICATION

Follow the directions for each application. Write the answer to each statement on the lines provided.

48. The four major tissue types in the body are epithelial, connective, muscle, and nervous. They are the components used to make organs. For each organ listed, give the tissue type that makes up most of the organ.

   a. Stomach

   b. Heart

   c. Spinal cord

   d. Bone

case **studies**

Write your response to each case study question on the lines provided.

49. A young man is using illicit drugs. How may this affect homeostasis?

50. A 52-year-old construction worker was impaled on a metal rod that entered his body just below his navel. If it went completely through his body from anterior to posterior, what body cavities did the metal rod pass through?

___

pathophysiology

Write your response to the question on the lines provided.

51. How does knowing normal anatomy and physiology help you understand pathophysiology?

Concepts of Chemistry

vocabulary review

MATCHING

Match the key terms in the right column with the definitions in the left column by writing the letter of the correct answer in the space provided.

_____ 1. Anything that takes up space

_____ 2. Neutrally charged particles found in the center of an atom

_____ 3. Negatively charged particles surrounding the nucleus

_____ 4. The number of protons in an element

_____ 5. Contains the genetic information of the cell

_____ 6. Combination of two or more different atoms of different elements

_____ 7. Unit of matter that makes up a chemical element

_____ 8. A branch of chemistry dealing with the chemistry of life

_____ 9. What holds atoms together

_____ 10. Small molecules combine to form larger molecules

_____ 11. The breakdown of complex molecules into simpler molecules with the release of energy

_____ 12. Atoms with the same atomic number, but different atomic weights

_____ 13. An atom or group of atoms with a positive charge

_____ 14. Used to store energy for cells

_____ 15. The center of the atom where protons and neutrons are located

a. Anabolism
b. Anion
c. Atom
d. Atomic number
e. Atomic weight
f. Biochemistry
g. Catabolism
h. Cation
i. Chemical bond
j. Compound
k. DNA
l. Electrons
m. Isotopes
n. Matter
o. Neutrons
p. Nucleus
q. Protons
r. Ribosomes
s. RNA
t. Triglycerides

content review

MULTIPLE CHOICE

In the space provided, write the letter of the choice that best completes each statement or answers each question.

_____ 16. What is formed when two or more atoms of one or more elements are combined?
   a. Compound
   b. Cation
   c. Anion
   d. Polyion

_____ 17. What compound is composed of two elements—hydrogen and oxygen?
   a. Water
   b. Air
   c. Carbon dioxide
   d. Sugar

_____ 18. What is the sum of all chemical reactions that take place in the human body?
   a. Metabolism
   b. Anabolism
   c. Catabolism
   d. Hydrolysis

_____ 19. What are the two processes of metabolism?
   a. Anabolism and catabolism
   b. Synthesis and degradation
   c. Positive and negative feedback
   d. Synthesis and feedback

_____ 20. What can be gas, liquid, or solid?
   a. Energy of motion
   b. Kinetic energy
   c. Potential energy
   d. Matter

_____ 21. What term is used interchangeably with *atom*?
   a. Element
   b. Proton
   c. Neutron
   d. Electron

_____ 22. What are neutrally charged particles in a nucleus called?
   a. Neutrons
   b. Particles
   c. Electrons
   d. Protons

_____ 23. What are positively charged particles in a nucleus called?
   a. Electrons
   b. Neutrons
   c. Protons
   d. Positrons

_____ 24. Matter is divided into what two large categories?
   a. Organic and synthetic
   b. Anabolic and catabolic
   c. Synthetic and hydrolytic
   d. Organic and inorganic

_____ 25. What is the difference between an organic and an inorganic molecule?
   a. Organic matter forms larger molecules and contains carbon and hydrogen.
   b. Inorganic matter forms larger molecules and contains carbon and hydrogen.
   c. Organic matter forms smaller molecules and contains carbon and hydrogen.
   d. Inorganic matter forms smaller molecules and contains carbon and hydrogen.

_____ 26. Examples of inorganic substances are
   a. Carbohydrates, water, and salts
   b. Water, oxygen, carbon dioxide, and salts
   c. Proteins and water
   d. Carbon dioxide and fats

UNIT 1   The Human Body and Disease

_____ 27. What inorganic substance regulates the body's temperature?
   a. Water
   b. HCl
   c. NaCl
   d. Carbon dioxide

_____ 28. What essential inorganic molecule is inhaled?
   a. Carbon dioxide
   b. Carbon monoxide
   c. Oxygen
   d. Chloride

_____ 29. What is exchanged in the cells for oxygen?
   a. Carbon dioxide
   b. Carbon monoxide
   c. Nitrogen
   d. Water

_____ 30. What are the four major classes of organic matter in the body?
   a. Carbohydrates, triglycerides, proteins, and nucleic acids
   b. Carbohydrates, lipids, proteins, and nucleic acids
   c. Carbohydrates, fats, DNA, and RNA
   d. Water, oxygen, carbon dioxide, and salts

_____ 31. What organic matter is better known as sugar?
   a. Carbohydrates
   b. Proteins
   c. Lipids
   d. Nucleic acids

_____ 32. What is an example of nucleic acid?
   a. DNA
   b. Protein
   c. TNA
   d. BNA

_____ 33. Genetic information is stored in
   a. DNA
   b. RNA
   c. Proteins
   d. Lipids

_____ 34. What is the number of protons in the element?
   a. Atomic mass
   b. Atomic weight
   c. Atomic number
   d. Atomic nomenclature

_____ 35. A sodium ion and a chloride ion attracted to each other form
   a. Sodium chloride
   b. Hydrogen peroxide
   c. Sodium nitrate
   d. A base

FILL IN THE BLANKS

In the space provided, write the word or phrase that best completes each sentence. Not all words or phrases are used.

36. Deoxyribonucleic acid is also more commonly known as _____.

37. _____ is a branch of chemistry that deals with the chemistry of life.

38. _____ is stored in chemical bonds. When the bonds are broken, _____ is released into the body.

39. Two oxygen atoms and one carbon atom sharing _____ form carbon dioxide.

40. Matter can be either _____, _____, or _____.

41. The atomic number of an element is the number of _____ in the element.

42. Isotopes are atoms with the _____ atomic number, but _____ atomic weights.

43. _____ matter generally does not contain carbon and hydrogen.

44. Water can prevent the pH of our body from moving too far from a _____ range.

45. We inhale oxygen and then it is transported through our blood attached to _____.

a. Biochemistry
b. Different
c. DNA
d. Electrons
e. Energy
f. Gas
g. Hemoglobin
h. Inorganic
i. Liquid
j. Neutrons
k. Normal
l. Protons
m. RNA
n. Same
o. Solid

SHORT ANSWER

Write the answer to each question or statement on the lines provided.

46. Explain the transportation of oxygen to cells and tissues in the body.

_____

_____

_____

47. Identify the four major classes of organic matter.

48. Explain the similarities and differences between DNA and RNA.

49. Explain the function of water in the body.

50. How does a chemical reaction impact life span?

LABELING

Follow the directions and write the answers on the lines provided.

51. Identify these terms in the following figure by writing them on the lines provided: *covalent bond, electron, neutron.*

a. _____

b. _____

c. _____

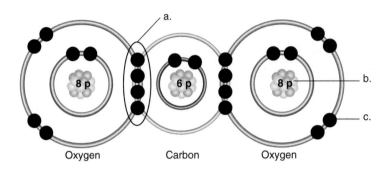

critical **thinking** / application

CRITICAL THINKING

Write the answer to the question on the lines provided.

52. Why is cholesterol essential to life?

_____
_____
_____
_____

APPLICATION

Follow the directions for each application. Write the answer to the question on the lines provided.

53. How are isotopes used in medicine?

_____
_____
_____
_____

case studies

Write your response to the case study statement on the lines provided.

54. A pregnant woman has heard about PKU, but does not know what it is. Explain what the disorder is and how it may be prevented.

_____
_____
_____
_____

pathophysiology

Write your response to the question on the lines provided.

55. How does knowing chemistry and biochemistry help you understand pathophysiology?

_____
_____
_____
_____

Concepts of Cells and Tissues 3

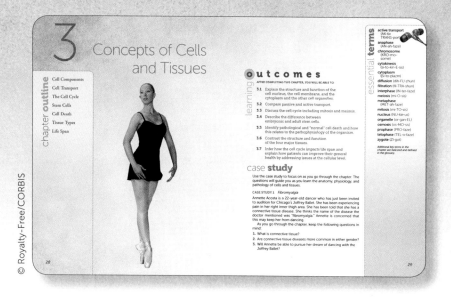

vocabulary review

MATCHING

Match the key terms in the right column with the definitions in the left column by writing the letter of the correct answer in the space provided.

_____ 1. An organelle that contains digestive enzymes

_____ 2. The division of the cell nucleus that results in two identical nuclei

_____ 3. The final stage of mitosis

_____ 4. The phase of mitosis where the nuclear envelope breaks up and chromatin condenses to form chromosomes

_____ 5. Distribution of cytoplasm into two separate cells during cell division

_____ 6. The period between cell divisions when the cell is involved in growth and various metabolic processes

_____ 7. The movement of substances across cell membranes against a concentration gradient and requiring energy

_____ 8. A structure found within the nucleus of the cell made up of DNA

_____ 9. The net movement of water molecules through a selectively permeable membrane

_____ 10. A small structure within the cell that has a specific function.

_____ 11. Movement of molecules or ions from an area of high concentration to an area of low concentration and requiring no energy

_____ 12. A type of cell division that takes place during gamete production

_____ 13. The flow of a liquid through a filter due to hydrostatic pressure

_____ 14. The stage of mitosis where spindle fibers shorten, dragging chromatids to opposite ends of the cell

_____ 15. Stage of mitosis where chromatids line up in the center of the cell

a. Active transport
b. Anaphase
c. Chromosome
d. Cytokinesis
e. Diffusion
f. Filtration
g. Interphase
h. Lysosome
i. Meiosis
j. Metaphase
k. Mitosis
l. Organelle
m. Osmosis
n. Prophase
o. Telophase

content review

MULTIPLE CHOICE

In the space provided, write the letter of the choice that best completes each statement or answers each question.

_____ 16. Which of the following is the basic unit of life?
  a. Cell
  b. Nucleus
  c. Mitochondria
  d. Gene

_____ 17. The study of tissues is called
  a. Histology
  b. Cytology
  c. Microbiology
  d. Anatomy

_____ 18. The process of cells becoming specialized is called
  a. Differentiation
  b. Speciation
  c. Dedifferentiation
  d. Hybridization

_____ 19. The fertilized egg that forms from the union of a sperm and an ovum is called a(n)
  a. Zygote
  b. Spermatid
  c. Oocyte
  d. Morula

_____ 20. What are the three components of a typical mature human body cell?
  a. Cytosol, cytoplasm, and cell membrane
  b. Nucleus, cytosol, and cytoplasm
  c. Nucleus, cytoplasm, and cell membrane
  d. Nucleus, nucleolus, and cytoplasm

_____ 21. A mature red blood cell is called a(n)
  a. Thrombocyte
  b. Leukocyte
  c. Platelet
  d. Erythrocyte

_____ 22. What is sometimes called the "brain" of the cell?
  a. Gene
  b. Mitochondrion
  c. Nucleus
  d. Chromosome

_____ 23. The proteins in the nucleus of a cell are called
  a. Histones
  b. Ribosomes
  c. Proteases
  d. Collagens

_____ 24. Histones hold together condensed chromatin or
  a. Myosomes
  b. Chromosomes
  c. Desmosomes
  d. Sclerotomes

_____ 25. What is the fluid environment within the nucleus called?
  a. Nucleoplasm
  b. Cytoplasm
  c. Cytosol
  d. Amorphic organelles

_____ 26. What is the portion of a cell between the cell membrane and the nucleus?
  a. Cytosol
  b. Cytoplasm
  c. Nucleolus
  d. Cell wall

_____ 27. What besides cytosol can be found in cytoplasm?
  a. Organelles
  b. Cell membrane
  c. Plasma membrane
  d. Histones

_____ 28. A mitochondrion is called the
   a. Powerhouse of the cell
   b. Brain of the cell
   c. Protein-making factory
   d. Channeling system of the cell

_____ 29. Mature red blood cells carry how many mitochondria?
   a. Between 1,000 and 20,000 per red blood cell
   b. More than 10,000 per red blood cell
   c. Very little
   d. None

_____ 30. The inner membrane folds of mitochondria are called
   a. Cristae
   b. Plasma membrane
   c. Nuclear membrane
   d. Nuclear envelope

_____ 31. Ribosomes are made up of
   a. Ribosomal RNA
   b. Transfer RNA
   c. Messenger RNA
   d. RNA and DNA

_____ 32. A network of channels that has a single membrane is called the
   a. Nuclear reticulum (ER)
   b. Plasma reticulum
   c. Endoplasmic reticulum
   d. Cytoplasmic reticulum

_____ 33. Two forms of ER are
   a. mER and sER
   b. rER and mER
   c. rER and sER
   d. mEr and tER

_____ 34. What hairlike projections are on the outside of the cell membrane and help move substances along the cell surface?
   a. Cilia
   b. Flagella
   c. Villi
   d. Microvilli

_____ 35. What hairlike projections are found only on sperm cells and are used for locomotion of the cell?
   a. Villi
   b. Flagella
   c. Microvilli
   d. Cilia

_____ 36. What is the movement of molecules or ions from an area of higher concentration to an area of lower concentration called?
   a. Diffusion
   b. Active transport
   c. Massive transport
   d. Mass effect

FILL IN THE BLANKS

In the space provided, write the word or phrase that best completes each sentence. Not all words or phrases are used.

37. A fertilized egg or zygote results from _____.

38. _____ is the genetic material in the cell.

39. All cells except gametes have 23 pairs or 46 _____.

40. Organelles are small permanent structures within the cytoplasm that serve specific _____.

41. _____, the inside of the cell, is made of cytosol and organelles.

42. ATP is called the _____ of the cell.

43. A single-stranded nucleic acid made up of ribose, nitrogenous bases, and phosphate group is called _____.

44. The _____ is a system of six or so stacked membrane sacs that packages and processes substances.

45. _____ are small, membrane-bound sacs containing hydrogen peroxide that kill foreign organisms.

46. Cartilage is a rigid _____.

47. Molecules such as glucose, potassium, and sodium enter cells through _____.

a. Active transport
b. Chromosomes
c. Connective tissue
d. Cytoplasm
e. DNA
f. Endoplasmic reticulum
g. Energy currency
h. Epithelium
i. Facilitated diffusion
j. Functions
k. Golgi apparatus
l. Peroxisomes
m. RNA
n. Structure
o. The union of a sperm and an ovum

SHORT ANSWER

Write the answer to each question or statement on the lines provided.

48. What characteristics distinguish a mature cell from an immature cell?

_____

_____

_____

49. List the typical mature cell components.

50. Explain the function of DNA in a cell.

51. Explain how ATP is produced and in what cells it can be found.

52. Describe how the cilia operate in the respiratory system.

LABELING

Follow the directions and write the answers on the lines provided.

53. Identify these terms in the following figure by writing them on the lines provided: *cell membrane, cytoplasm, nucleus.*

a. _____

b. _____

c. _____

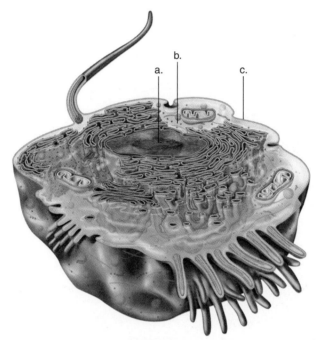

From *Hole's Human Anatomy & Physiology,* 12e, by Shier/Butler/and Lewis.
Copyright © 2009. Reprinted by permission of McGraw-Hill Companies Inc.

54. Identify these terms in the following figure by writing them on the lines provided: *centrioles, chromosomes, daughter cells, nuclear envelope.*

a. _____

b. _____

c. _____

d. _____

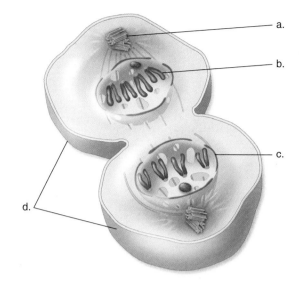

55. Identify these terms in the following figure by writing them on the lines provided, showing the tissue's ability to regenerate: *labile, permanent, stable*.

   a. _____

   b. _____

   c. _____

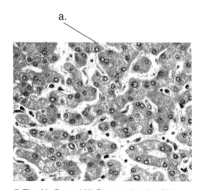

© The McGraw-Hill Companies, Inc./Al Telser, photographer

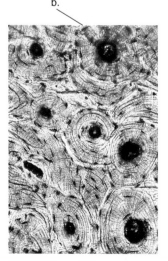

© The McGraw-Hill Companies, Inc./Dennis Strete, photographer

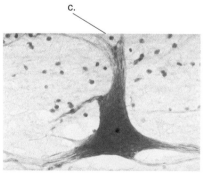

© Allen Bell/Corbis RF

CHAPTER 3   Concepts of Cells and Tissues

critical thinking / application

CRITICAL THINKING

Write the answer to each question on the lines provided.

56. Why is it necessary for gametes to undergo meiosis?

57. Why is the mature red blood cell void of a nucleus and other organelles that are commonly found in other cells?

APPLICATION

Follow the directions for each application. Write the answer to the question on the lines provided.

58. How may understanding the concept of apoptosis help with cancer research?

case studies

Write your response to the case study question on the lines provided.

59. A 22-year-old man has thrombocytopenia. What is this and how are homeostasis and his health affected by this condition?

pathophysiology

Write your response to the question on the lines provided.

60. How does knowledge of cells and tissues help you understand fibromyalgia?

Concepts of Disease 4

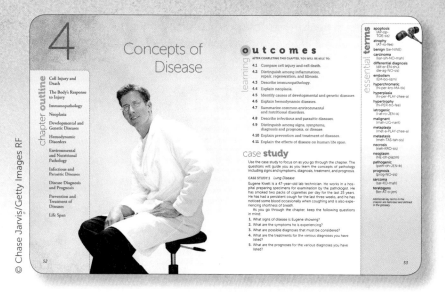

vocabulary review

MATCHING

Match the key terms in the right column with the definitions in the left column by writing the letter of the correct answer in the space provided.

_____ 1. A reduction in the size of a tissue or organ

_____ 2. An increase in the size of a tissue or organ not due to an increase in the number of cells

_____ 3. The change from one adult cell type to another adult cell type

_____ 4. "Programmed" or "suicidal" cell death

_____ 5. Pathological cell death

_____ 6. A cancer of epithelial origin

_____ 7. A "new" growth that may be benign or malignant

_____ 8. The ability of a tumor to spread to a distant site from its origin

_____ 9. Unintended injuries caused by physicians or prescribed drugs and therapies

_____ 10. An organism or substance capable of causing a birth defect in a fetus

_____ 11. A cancer of mesenchymal origin

_____ 12. A tumor that stays localized and does not metastasize

_____ 13. The outcome of a disease process

_____ 14. The movement of any substance through the arterial or venous circulations

_____ 15. Disordered growth and maturation of cells

a. Apoptosis
b. Atrophy
c. Benign
d. Carcinoma
e. Dysplasia
f. Embolism
g. Hyperplasia
h. Hypertrophy
i. Iatrogenic
j. Malignant
k. Metaplasia
l. Metastasis
m. Necrosis
n. Neoplasm
o. Prognosis
p. Sarcoma
q. Teratogen

content review

MULTIPLE CHOICE

In the space provided, write the letter of the choice that best completes each statement or answers each question.

_____ 16. Any structural or physiological change that disrupts homeostasis is called
   a. Disease
   b. Anatomy
   c. Necrosis
   d. Ambiosis

_____ 17. Which of the following is a way for a cell to adapt to injury?
   a. Inflammation
   b. Fibrosis
   c. Regeneration
   d. Hypertrophy

_____ 18. As we get older, we lose body mass. What is the term for this process?
   a. Atrophy
   b. Metaplasia
   c. Hypertrophy
   d. Hyperplasia

_____ 19. Endometrium producing more cells in response to estrogen is called
   a. Necrosis
   b. Hypertrophy
   c. Metaplasia
   d. Hyperplasia

_____ 20. Columnar epithelium changing to squamous epithelium due to chronic irritation is referred to as
   a. Metaplasia
   b. Hyperplasia
   c. Necrosis
   d. Apoptosis

_____ 21. What is a common condition in diabetics and alcoholics that results from an increase in intracellular storage of fat?
   a. Hepatocellular carcinoma
   b. Cirrhosis of the liver
   c. Fatty liver
   d. Hepatitis C

_____ 22. Which of the following is considered a precancerous condition?
   a. Dysplasia
   b. Hyperplasia
   c. Necrosis
   d. Apoptosis

_____ 23. Which of the following is associated with aging?
   a. Lipofuschin
   b. Glycogen
   c. Melanin
   d. Glucose

_____ 24. All of the following are signs of acute inflammation *except*
   a. Calor
   b. Pallor
   c. Tumor
   d. Dolor

_____ 25. When a cell is unable to adapt after injury, _____ may result.
   a. Necrosis
   b. Apoptosis
   c. Metaplasia
   d. Hyperplasia

_____ 26. Scarring is a type of
   a. Hypertrophy
   b. Fibrosis
   c. Metaplasia
   d. Hyperplasia

_____ 27. What is the term for excessive scar formation?
   a. Keloid
   b. Myeloid
   c. Apoptosis
   d. Cirrhosis

_____ 28. AIDS is the result of what type of infection?
   a. Fungal
   b. Bacterial
   c. Viral
   d. Parasitic

_____ 29. HIV is spread through all of the following *except*
   a. Semen
   b. Blood
   c. Shaking hands
   d. Breast milk

_____ 30. HIV targets
   a. T cells
   b. B cells
   c. Erythrocytes
   d. Adipocytes

_____ 31. What type of cancer of the connective tissue around blood vessels is seen in AIDS patients?
   a. Kaposi sarcoma
   b. Leukemia
   c. Iron deficiency anemia
   d. Melanoma

_____ 32. All of the following are examples of autoimmune disease *except*
   a. SLE
   b. Rheumatoid arthritis
   c. Sjögren syndrome
   d. HIV

_____ 33. _____ is a disease affecting the blood vessels and collagen of the skin and various organs.
   a. Scleroderma
   b. HIV
   c. Melanoma
   d. Leukemia

_____ 34. Breast cancers are classified as
   a. Carcinomas
   b. Sarcomas
   c. Leukemias
   d. Melanomas

_____ 35. Which of the following is used to determine how far a cancer has spread?
   a. Grading
   b. Staging
   c. Metastasis
   d. Invasion

CHAPTER 4   Concepts of Disease   29

FILL IN THE BLANKS

In the space provided, write the word or phrase that best completes each sentence. Not all words or phrases are used.

36. _____ is a spectrum of physical and mental disabilities and malformations as the result of intrauterine exposure of the fetus to alcohol.

37. _____ are any chemical, physical, or biological agents that can cause birth defects.

38. Exposure to rubella during the _____ trimester of pregnancy can cause birth defects in the fetus.

39. _____ are disorders that are passed on through the sex chromosomes.

40. In _____, both genetic and environmental factors play a role in the onset of the disease.

41. An aggregate of coagulated blood containing platelets, fibrin, and other elements within a blood vessel is known as a(n) _____.

42. _____ is another name for a heart attack.

43. _____ is an accumulation of excess fluid between cells.

44. _____ is an extreme hypersensitivity reaction that leads to widespread vasodilation and may result in death.

45. _____ is the single largest factor for preventable deaths in the United States.

a. Alcoholism
b. Anaphylaxis
c. Cerebrovascular accident
d. Edema
e. Fetal alcohol syndrome
f. First
g. Multifactorial inheritance
h. Myocardial infarction
i. Second
j. Sex-linked disorders
k. Smoking
l. Teratogens
m. Thrombus
n. Y-linked disorders

SHORT ANSWER

Write the answer to each statement on the lines provided.

46. Compare cell injury and cell death.

_____

_____

_____

_____

47. Compare acute and chronic inflammation.

48. Explain neoplasia.

49. Distinguish between signs and symptoms.

50. Distinguish among repair, regeneration, and fibrosis.

LABELING

Follow the directions and write the answers on the lines provided.

51. What are the organs shown in the following figure? Which one is abnormal?

_____

_____

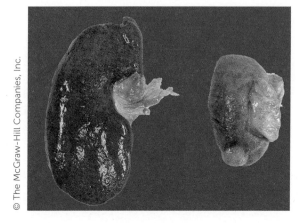

52. What is the organ shown below and what is the abnormality?

_____

_____

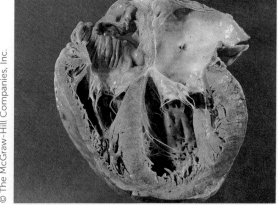

critical thinking / application

CRITICAL THINKING

Write the answer to the statement on the lines provided.

53. Explain why regeneration is not common in humans.

_____

_____

_____

_____

APPLICATION

Write the answer to the statement on the lines provided.

54. Explain why a thorough patient medical history is essential to making an accurate diagnosis.

_____

_____

_____

_____

case studies

Write your response to each case study question on the lines provided.

55. A 45-year-old stockbroker has been diagnosed with GERD. What are some complications of this disorder?

_____

_____

_____

_____

56. A young woman has been told that her mother is a carrier for hemophilia. What is the likelihood that the young woman is also a carrier?

pathophysiology

Write your response to the question on the lines provided.

57. What organs are affected by alcoholism and what is the treatment for this disorder?

Concepts of Microbiology

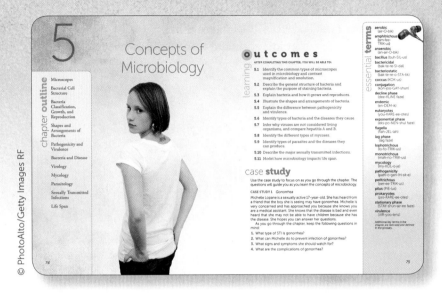

vocabulary review

MATCHING

Match the key terms in the right column with the definitions in the left column by writing the letter of the correct answer in the space provided.

_____ 1. An organism that does not have a nucleus

_____ 2. Having flagella at opposite ends of a bacterial cell

_____ 3. The last phase of the growth curve of bacteria caused by the death of cells exceeding the production of cells

_____ 4. Transmission of genetic material from one bacterium to a second

_____ 5. The first phase of the bacterial growth curve in which there is no evident growth in numbers of bacteria

_____ 6. Phase of the bacterial growth curve where there is very rapid growth

_____ 7. An organism whose cells have a nucleus

_____ 8. Having multiple flagella on one end of a bacterial cell

_____ 9. Unable to kill an organism, but able to inhibit its multiplication

_____ 10. A bacterium that is longer than it is wide

_____ 11. The degree of pathogenicity of an organism that has the ability to cause disease

_____ 12. The study of fungi

_____ 13. Organism that cannot survive in oxygen

_____ 14. Very common to a region or area

_____ 15. Able to kill bacteria

a. Aerobic
b. Amphitrichous
c. Anaerobic
d. Bacillus
e. Bactericidal
f. Bacteriostatic
g. Conjugation
h. Decline phase
i. Endemic
j. Eukaryote
k. Exponential phase
l. Flagellum
m. Lag phase
n. Lophotrichous
o. Mycology
p. Pilus
q. Prokaryote
r. Stationary phase
s. Virulence

35

content review

MULTIPLE CHOICE

In the space provided, write the letter of the choice that best completes each statement or answers each question.

_____ 16. The ability to distinguish two objects as being separate and distinct objects is called
   a. Identification
   b. Resolution
   c. Magnification
   d. Illumination

_____ 17. Which of the following microscopes is routinely used to identify *T. pallidum*?
   a. Darkfield microscope
   b. Light microscope
   c. SEM
   d. TEM

_____ 18. Which structure is used for motility of the organism?
   a. Cilia
   b. Pilus
   c. Flagellum
   d. Microvilli

_____ 19. Which structure is used for attachment to other cells as well as transmission of genetic material from one bacterium to another?
   a. Microvilli
   b. Pilus
   c. Flagellum
   d. Villi

_____ 20. Binomial nomenclature involves
   a. Kingdom and phylum names
   b. Genus and family names
   c. Genus and species names
   d. Species name only

_____ 21. What is the third phase of the bacterial growth curve called, where growth and death of the bacteria are balanced?
   a. Exponential phase
   b. Decline phase
   c. Logarithmic phase
   d. Stationary phase

_____ 22. Which type of drug can kill infectious organisms?
   a. Bacteriostatic
   b. Microcytic
   c. Bactericidal
   d. Macrocytic

_____ 23. A curved, comma-shaped bacterium is called a
   a. Spirochete
   b. Coccus
   c. Vibrio
   d. Bacillus

_____ 24. The ability of an organism to cause disease is referred to as
   a. Virulence
   b. Pathogenicity
   c. Contagiousness
   d. inflammation

_____ 25. All of the following are true of *C. perfringens* except
   a. It causes a type of gangrene
   b. Treatment may include antitoxins and antibiotics
   c. It is a Gram-negative rod
   d. It is an anaerobic

36   UNIT 1   The Human Body and Disease

_____ 26. Trismus is associated with
   a. *C. tetani*
   b. *S. aureus*
   c. *S. pyogenes*
   d. *T. pallidum*

_____ 27. The toxin of _____ can cause food poisoning.
   a. *C. botulinum*
   b. *S. pyogenes*
   c. *T. pallidum*
   d. *N. gonorrhea*

_____ 28. All of the following are true of *S. aureus* except
   a. The organism is arranged in pairs
   b. It is salt tolerant
   c. Staphylococcal toxins are resistant to heat
   d. It causes food poisoning

_____ 29. All of the following are true of enterobacteria *except*
   a. Some members of this group are part of the normal flora of humans
   b. They commonly infect the respiratory system
   c. They are very common pathogenic organisms
   d. *E. coli* belongs to this group of bacteria

_____ 30. Which of the following organisms is associated with duodenal ulcers?
   a. *H. pylori*
   b. *S. aureus*
   c. *T. pallidum*
   d. *M. tuberculosis*

_____ 31. Which of the following are *not* considered living organisms?
   a. Viruses
   b. Bacteria
   c. Fungi
   d. Parasites

_____ 32. Which of the following is *not* a virus?
   a. *T. pallidum*
   b. Hepatitis A
   c. Hepatitis C
   d. HIV

_____ 33. What is the term for the study of fungi?
   a. Mycology
   b. Myology
   c. Virology
   d. Biology

_____ 34. Black piedra is a _____ infection.
   a. Bacterial
   b. Fungal
   c. Viral
   d. Helminthic

_____ 35. Which of the following is *not* a category of parasite?
   a. Cocci
   b. Helminthes
   c. Protozoa
   d. Arthropods

FILL IN THE BLANKS

In the space provided, write the word or phrase that best completes each sentence. Not all words or phrases are used.

36. _____ is a type of yeast infection that is commonly found in moist areas of the body such as the oral mucosa, vagina, and axilla.

37. When people are infected with parasites, there is typically an increase in _____, a type of white blood cell.

38. Approximately 90 percent of cervical cancers are caused by _____.

39. A chancre is the lesion typically seen with _____.

40. Syphilis is a(n) _____ infection.

41. _____ is known as genital warts.

42. A(n) _____ is an organism that needs a host to survive.

43. _____ is contracted through the inhalation of dried bird droppings, especially pigeons and chickens.

44. Individuals with _____ immune systems are at increased risk for opportunistic infections.

45. Mycobacteria are identified through the _____ stain.

a. Acid fast
b. Bacterial
c. *C. albicans*
d. Compromised
e. *C. acuminatum*
f. Eosinophils
g. Gonorrhea
h. Gram varicella
i. Histoplasmosis
j. Human papillomavirus
k. Parasite
l. Syphilis
m. Viral

SHORT ANSWER

Write the answer to each statement on the lines provided.

46. Explain the bacterial growth curve.

_____

_____

_____

_____

47. Infer the importance of normal flora to homeostasis.

48. Describe syphilis and its stages.

49. Explain how microbiology impacts life span.

50. Describe the shapes and arrangement of bacteria.

LABELING

Follow the directions and write the answers on the lines provided.

51. Identify these terms in the bacterial growth curve shown below by writing them on the lines provided: *decline phase, exponential phase, lag phase,* and *stationary phase.*

a. _____

b. _____

c. _____

d. _____

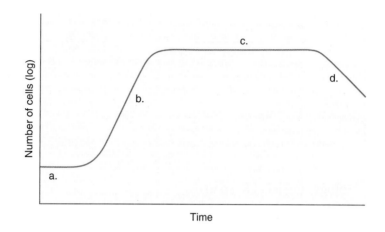

52. Indicate the correct flagella arrangement in the following figure using these terms and writing them on the lines provided: *amphitrichous, lophotrichous, monotrichous, peritrichous.*

a. _____

b. _____

c. _____

d. _____

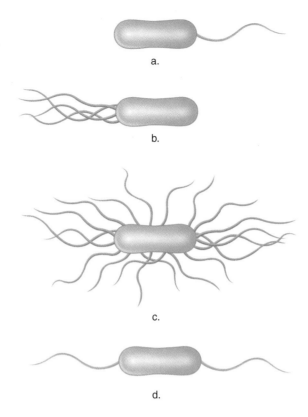

a.

b.

c.

d.

critical thinking / application

CRITICAL THINKING

Write the answer to each statement on the lines provided.

53. Indicate why someone infected with syphilis must take the entire regimen of antibiotics even if he or she seems to be better and why the individual must notify all his or her sexual partners.

_____
_____
_____
_____

54. Explain why someone placed on a broad spectrum antibiotic that cleanses the gut of normal flora may be at increased risk for bleeding tendencies.

APPLICATION

Follow the directions for the application. Write the answer to the question on the lines provided.

55. How does Gram staining help determine the type of antibiotic that may be prescribed for an infection?

case studies

Write your response to the case study question on the lines provided.

56. A health care professional accidently stuck herself while disposing of a needle recently used on a patient. What are some infections she is at risk of contracting and what precautions should she take at this point?

pathophysiology

Follow the instructions for the question.

57. Fill in the missing cells.

Disease	Etiology	Signs and Symptoms	Treatment
Tetanus			
Tuberculosis			
Ringworm			
Gonorrhea			

Concepts of Fluid, Electrolyte, and Acid–Base Balance

vocabulary review

MATCHING

Match the key terms in the right column with the definitions in the left column by writing the letter of the correct answer in the space provided.

_____ 1. A hormone secreted by the right atrium to increase sodium and water excretion

_____ 2. Substance that dissolves in water and is capable of carrying an electrical current

_____ 3. Condition in which the plasma pH is less than 7.35

_____ 4. High potassium levels in the blood

_____ 5. Can convert strong acids and bases to weak acids and bases, respectively

_____ 6. A plasma potassium concentration less than normal

_____ 7. The fluid found between cells

_____ 8. A hormone that causes vasoconstriction and causes the adrenal cortex to secrete aldosterone

_____ 9. Another name for ADH

_____ 10. A higher than normal plasma concentration of chloride

_____ 11. A lower than normal plasma concentration of phosphate

_____ 12. A hormone secreted by the adrenal cortex for the absorption of sodium by the kidneys

_____ 13. Condition in which the plasma pH is greater than 7.45

_____ 14. Acts a buffer to raise the plasma pH

_____ 15. High sodium levels in the blood

a. Acidosis
b. Aldosterone
c. Alkalosis
d. Angiotensin II
e. Atrial natriuretic peptide
f. Bicarbonate ion
g. Buffer
h. Electrolyte
i. Hyperchloremia
j. Hyperkalemia
k. Hypernatremia
l. Hypokalemia
m. Hypophosphatemia
n. Interstitial fluid
o. Vasopressin

content review

MULTIPLE CHOICE

In the space provided, write the letter of the choice that best completes each statement or answers each question.

_____ 16. The amount of water in an adult male is about _____ of the total body mass.
   a. 40 percent
   b. 20 percent
   c. 60 percent
   d. 80 percent

_____ 17. The majority of the total body water is found in what substance?
   a. Intracellular fluid
   b. Interstitial fluid
   c. Plasma
   d. Edema

_____ 18. What is the term for the liquid component of the cytoplasm?
   a. Plasma
   b. Cytosol
   c. Interstitial fluid
   d. Intrastitial fluid

_____ 19. About _____ of our daily water gain is from water and other liquids that we drink.
   a. 1,600 mL
   b. 400 mL
   c. 600 mL
   d. 1,000 mL

_____ 20. The food we eat supplies about _____ of our daily water intake.
   a. 1000 mL
   b. 200 mL
   c. 600 mL
   d. 1,600 mL

_____ 21. The hormone _____ is released from the posterior pituitary in response to low fluid volume in the body.
   a. Aldosterone
   b. Vasopressin
   c. Oxytocin
   d. Growth hormone

_____ 22. A positively charged ion is called a(n)
   a. Anion
   b. Neutron
   c. Cation
   d. Positron

_____ 23. Which of the following accounts for 90 percent of the extracellular cations?
   a. Sodium
   b. Potassium
   c. Magnesium
   d. Hydrogen

_____ 24. Fluid loss that is not consciously perceived is called
   a. Perspiration
   b. Exhalation
   c. Micturition
   d. Insensible

_____ 25. If we ingest more salt, all of the following will happen *except*
   a. Swelling of tissues
   b. Drop in blood pressure
   c. Increased thirst
   d. Secretion of ANP

26. Increased secretion of ANP will
    a. Increase sodium excretion, but decrease water excretion
    b. Increase both sodium and water excretion
    c. Decrease water excretion, but increase sodium excretion
    d. Decrease both sodium and water excretion

27. Aldosterone is secreted by which of the following?
    a. Pituitary gland
    b. Adrenal cortex
    c. Adrenal medulla
    d. Kidneys

28. What is the term for substances that dissolve in solution and are capable of carrying an electrical current?
    a. Lipids
    b. Electrolytes
    c. Triglycerides
    d. Nucleic acids

29. The normal plasma concentration of the extracellular fluid is
    a. 100 mOsm/L
    b. 500 mOsm/L
    c. 300 mOsm/L
    d. 700 mOsm/L

30. Natrium is the chemical name for which of the following?
    a. Sodium
    b. Magnesium
    c. Hydrogen
    d. Calcium

31. Where is the majority of the body's calcium stored?
    a. Plasma
    b. The CNS
    c. Bones and teeth
    d. The PNS

32. Hyperparathyroidism can cause which of the following?
    a. Hypercalcemina
    b. Hypocalcemina
    c. Acromegaly
    d. Myxedema

33. The majority of phosphate in the body is found in the bones and teeth as part of
    a. Manganese phosphate salts
    b. Magnesium phosphate salts
    c. Phosphoric acid
    d. Calcium phosphate salts

34. What is the normal pH of the arterial blood?
    a. 7.35–7.45
    b. 6.35–6.45
    c. 8.35–8.45
    d. 9.35–9.45

35. All of the following are reasons for administering intravenous fluids *except*
    a. To transfuse blood
    b. To decrease the volume load on the heart
    c. To replace fluids and electrolytes
    d. To administer medications

CHAPTER 6   Concepts of Fluid, Electrolyte, and Acid–Base Balance

FILL IN THE BLANKS

In the space provided, write the word or phrase that best completes each sentence. Not all words or phrases are used.

36. _____ is a condition in which the plasma pH is greater than 7.45.

37. _____ is common in chronic obstructive lung diseases such as emphysema.

38. Because the kidneys of infants are not fully developed, they may not be able to efficiently excrete hydrogen ions resulting in _____.

39. The _____ fluid compartment is composed of interstitial fluid and plasma.

40. _____ systems maintain the pH of the body around a very narrow range to protect homeostasis.

41. As a percentage of total body mass, the amount of water in women is _____ than in men.

42. _____ is another name for urination.

43. The _____ is the liquid component of the cytoplasm.

44. The amount of water we gain and lose each day is about _____ liters.

45. The two mechanisms by which we gain water are ingestion and _____.

a. Acidosis
b. Alkalosis
c. Buffer
d. Cytosol
e. Edema
f. Extracellular
g. 5.0
h. Greater
i. Hematuria
j. Intracellular
k. Less
l. Metabolism
m. Micturition
n. 2.5

SHORT ANSWER

Write the answer to each question on the lines provided.

46. For each of the following, give the percentage of plasma in the fluid.

   a. Total body water _____

   b. Extracellular fluid _____

47. For each of the following, give the correct volume.

   a. The amount of water the body loses every day _____

   b. The amount of urine excreted by the kidneys each day _____

LABELING

Follow the directions and write the answers on the lines provided.

48. Identify these terms in the following figure by writing them on the lines provided: *interstitial fluid, lymph, plasma, transcellular fluid*

a. _____

b. _____

c. _____

d. _____

Total Body Water

Membranes of body cells — Intracellular fluid (63%)

a. —
b. — Extracellular fluid (37%)
c. —
d. —

From *Hole's Human Anatomy & Physiology*, 12e, by Shier/Butler/and Lewis. Copyright © 2009. Reprinted by permission of McGraw-Hill Companies Inc.

critical thinking / application

CRITICAL THINKING

Write the answer to the statement on the lines provided.

49. Explain how fluids, electrolytes, and acid–base balance impact life span.

_____

_____

_____

APPLICATION

Follow the directions for the application. Write the answers to each statement on the lines provided.

50. Explain how elderly individuals can protect themselves from dehydration on hot, humid days.

_____

_____

_____

case studies

Write your response to each case study question on the lines provided.

51. A young man has had severe gastroenteritis with vomiting and diarrhea for two days. What electrolytes may be affected and what are some signs and symptoms he may be experiencing?

_____

_____

_____

_____

52. A pregnant woman was told by her dentist that she has three new cavities. Is there a connection between her pregnancy and the dental caries (cavities)?

_____

_____

_____

_____

pathophysiology

Follow the instructions for the statement.

53. Fill in the missing cells.

Condition	Electrolyte	Causes	Signs and Symptoms
Hypokalemia			
Hypernatremia			
Hypophosphatemia			
Hyperchloremia			

UNIT 2 Concepts of Common Illness by System

7

The Integumentary System

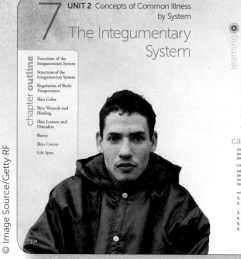

vocabulary review

MATCHING

Match the key terms in the right column with the definitions in the left column by writing the letter of the correct answer in the space provided.

_____ 1. Most superficial layer of the skin

_____ 2. Yellow pigment found in the stratum corneum of the epidermis

_____ 3. Hair loss (baldness)

_____ 4. Smooth muscle that is associated with a hair and "goose bumps"

_____ 5. Most numerous cell type in the epidermis

_____ 6. Epidermal cells that produce pigment

_____ 7. A wart

_____ 8. Skin disorder more common in children and caused by *S. aureus*

_____ 9. Blue discoloration of the skin, nails, or mucous membranes due to decreased oxygen in the blood

_____ 10. Layer of the epidermis found only in thick skin

a. Alopecia
b. Arrector pili
c. Carotene
d. Cyanosis
e. Epidermis
f. Erythema
g. Hyperdermis
h. Impetigo
i. Keratinocytes
j. Melanin
k. Melanocyte
l. Stratum corneum
m. Stratum lucidum
n. Verruca

51

content review

MULTIPLE CHOICE

In the space provided, write the letter of the choice that best completes each statement or answers each question.

_____ 11. Which of the following is responsible for the formation of fingerprints?
   a. Epidermis
   b. Dermis
   c. Hypodermis
   d. Subcutaneous

_____ 12. Which of the following are sudoriferous glands that produce response to high temperatures and act to cool the body?
   a. Apocrine
   b. Endocrine
   c. Eccrine
   d. Holocrine

_____ 13. Which area of skin is innervated by a single spinal nerve?
   a. Myotome
   b. Sclerotome
   c. Dermatome
   d. Monotome

_____ 14. Which of the following is adjacent to the dermis?
   a. Stratum basale
   b. Stratum corneum
   c. Stratum granulosum
   d. Stratum lucidum

_____ 15. Melanomas originate in which of the following skin layers?
   a. Epidermis
   b. Dermis
   c. Hypodermis
   d. Subcutaneous layer

_____ 16. Adipose tissue is most abundant in the
   a. Epidermis
   b. Dermis
   c. Hypodermis
   d. Stratum lucidum

_____ 17. Which of the following is necessary for the absorption of calcium?
   a. Vitamin A
   b. Vitamin B
   c. Vitamin C
   d. Vitamin D

_____ 18. Which type of skin cancer rarely metastasizes?
   a. Basal cell
   b. Squamous cell
   c. Melanoma
   d. Malignant melanoma

_____ 19. Which of the following is *not* a function of skin?
   a. Protection
   b. Vitamin A synthesis
   c. Sensation
   d. Body temperature regulation

_____ 20. Which of the following is *not* a determinant of skin coloration?
   a. Carotene
   b. Keratin
   c. Melanin
   d. Hemoglobin

21. The cause of chickenpox is _____ and the cause of shingles is _____.
    a. Viral, bacteria
    b. Bacterial, bacterial
    c. Bacterial, viral
    d. Viral, viral

22. Which inflammatory skin disorder is associated with excess sebum production?
    a. Acne vulgaris
    b. Herpes simplex 1
    c. Herpes simplex 2
    d. Psoriasis

23. What is another name for a decubitus ulcer?
    a. Duodenal ulcer
    b. Abrasion
    c. Avulsion
    d. Bedsore

24. What is another name for the stratum basale?
    a. Stratum germinativum
    b. Stratum lucidum
    c. Stratum corneum
    d. Stratum spinosum

25. A hair follicle is made up mostly of
    a. Keratinocytes
    b. Melanocytes
    c. Osteocytes
    d. Erythrocytes

26. Which of the following is an inflammation of the connective tissue in skin?
    a. Cellulitis
    b. Eczema
    c. Psoriasis
    d. Melanoma

27. Comedoes are associated with
    a. Acne vulgaris
    b. Impetigo
    c. Psoriasis
    d. Eczema

28. All of the following are true of the skin as aging occurs *except*
    a. Loss of collagen
    b. Loss of elasticity
    c. Loss of subcutaneous fat
    d. Increase in circulation to the skin

29. What is the term for folliculitis involving a single hair follicle?
    a. Furuncle
    b. Pimple
    c. Blister
    d. Carbuncle

30. A pus-filled lesion such as a pimple is called a(n)
    a. Pustule
    b. Excoriation
    c. Vesicle
    d. Bulla

FILL IN THE BLANKS

In the space provided, write the word or phrase that best completes each sentence. Not all words or phrases are used.

31. The skin entirely replaces itself every _____ days.

32. The _____ layer of the skin has a variety of receptors for light touch, pain, pressure, and temperature.

33. The most superficial layer of the epidermis is the _____.

34. The _____ is the crescent-shaped lighter area of the nail located distal to the cuticle.

35. When scarring is excessive, it may be called "proud flesh." The medical term for these types of scars is _____.

36. _____ are pinpoint hemorrhages on the skin or mucosa.

37. _____ is a fungal infection that is commonly referred to as ringworm.

38. A method of evaluating the severity of burns is the _____.

39. _____ uses freezing to kill cancer cells.

40. _____ gives skin and hair their coloring.

a. Carotene
b. Cryosurgery
c. Dermal
d. Follicle
e. 45
f. Keloids
g. Lunula
h. Melanin
i. Petechiae
j. Purpura
k. Rule of nines
l. Stratum corneum
m. 30 days
n. Tinea

SHORT ANSWER

Write the answer to each statement on the lines provided.

41. Describe the ABCs of melanoma.

_____

_____

_____

_____

54   UNIT 2   Concepts of Common Illness by System

42. Describe the characteristics of the layers of the epidermis.

_____

_____

_____

_____

_____

43. Compare and contrast the papillary and reticular regions of the dermis.

_____

_____

_____

_____

_____

_____

44. Differentiate between sudoriferous and sebaceous glands.

_____

_____

_____

_____

_____

_____

45. Describe the following skin disorders and lesions.

    a. Impetigo _____

    b. Psoriasis _____

    c. Acne _____

LABELING

Follow the directions and write the answers on the lines provided.

46. The following figure shows structures and the layers of the skin. Identify these terms in the figure by writing them on the lines provided: *adipose tissue, arrector pili muscle, basement membrane, blood vessels, capillary, dermal papilla, dermis, epidermis, hair shaft, hair follicle, Meissner's corpuscle, nerve cell process, sebaceous gland, stratum basale, stratum corneum, subcutaneous layer, sweat gland, sweat gland duct, sweat gland pore.*

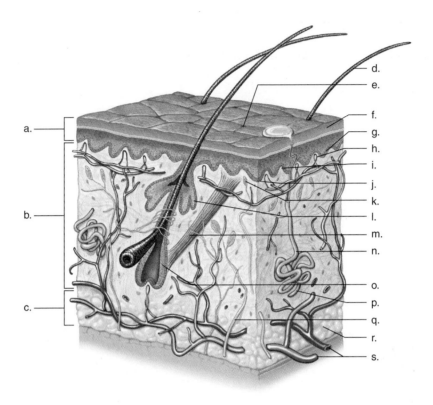

From *Hole's Human Anatomy & Physiology*, 12e, by Shier/Butler/and Lewis. Copyright © 2009. Reprinted by permission of McGraw-Hill Companies Inc.

a. _____
b. _____
c. _____
d. _____
e. _____
f. _____
g. _____
h. _____

56 UNIT 2 Concepts of Common Illness by System

i. _____

j. _____

k. _____

l. _____

m. _____

n. _____

o. _____

p. _____

q. _____

r. _____

s. _____

critical thinking / application

CRITICAL THINKING

Write the answer to each statement on the lines provided.

47. Describe how skin color is determined.

_____
_____
_____
_____

48. Explain the response of the circulatory system and the sudoriferous glands to high environmental temperatures.

_____
_____
_____
_____

APPLICATION

Follow the directions for the application.

49. Compare and contrast the types of skin cancer by filling in the missing cells. Include the risk factors, treatment, and the cells and layers of the skin that are first involved.

	Basal Cell	Squamous Cell	Melanoma
Risk factors			
Treatment			
Cells involved			

case studies

Write your response to each case study question on the lines provided.

50. A five-year-old boy is brought to the pediatrician's office by his parents. He has a "honey crusted" discharge around his nose. His parents say that a friend of his has something similar in appearance on his face as well. What condition or disease do you suspect the boy(s) of having? What is the cause and what is the treatment?

_____

_____

_____

_____

51. A 42-year-old female with fair skin and red hair has noticed a "mole" on her left shoulder. She says that she uses the tanning salon at least two or three times per week. Is this likely a mole or something more serious? What risk factors does she have? What is the treatment and prognosis if it is something more serious?

_____

_____

_____

_____

52. A 52-year-old man was burned in an industrial accident when he fell into some hot water that had spilled onto the floor in the foundry where he is a supervisor. He has third-degree burns on his right arm and back. He has second-degree burns on his right leg and right hand. What is the difference between first-, second-, and third-degree burns? Using the rule of nines, how would you evaluate him?

_____

_____

_____

_____

pathophysiology

Follow the instructions for the statement.

53. Fill in the missing cells.

Disease	Etiology	Signs/Symptoms	Treatment	Prognosis
Impetigo				
Acne vulgaris				
Eczema				

CHAPTER 7 The Integumentary System   59

The Skeletal System 8

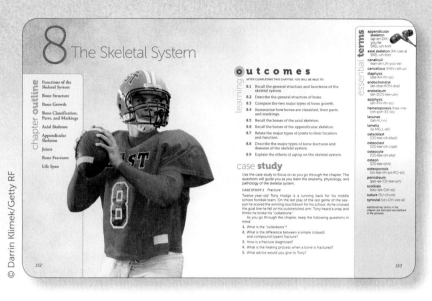

vocabulary review

MATCHING

Match the key terms in the right column with the definitions in the left column by writing the letter of the correct answer in the space provided.

_____ 1. Attaches the finger to the hand

_____ 2. Includes bones of the extremities

_____ 3. The opening for the spinal cord and brain attachment

_____ 4. Bone depression that allows for bones to join, as with the elbow

_____ 5. The baby's "soft spot"

_____ 6. The skeleton including the cranium, vertebrae, and rib cage

_____ 7. The term used to describe the finger joints

_____ 8. Nonmovable joints, such as those within the skull

_____ 9. Large, bony projection, such as at the proximal end of the femur

_____ 10. Attaches the lower extremities to the axial skeleton

_____ 11. Ridgelike bony projection

_____ 12. Also called the "breastbone"

_____ 13. The location of red bone marrow

_____ 14. A rounded bone process that usually articulates with another bone

_____ 15. A large, rounded process at the end of a long bone

a. Appendicular
b. Axial
c. Cancellous bone
d. Condyle
e. Crest
f. Fontanelle
g. Foramen magnum
h. Fossa
i. Head
j. Hyoid
k. MCP joints
l. Pelvic girdle
m. PIP/DIP
n. Sternum
o. Suture
p. Trochanter

61

content review

MULTIPLE CHOICE

In the space provided, write the letter of the choice that best completes each statement or answers each question.

_____ 16. Which of the following is an example of short bones?
   a. Carpal bones
   b. Humerus
   c. Ribs
   d. Sternum

_____ 17. The growth plate is also known as the
   a. Articular cartilage
   b. Diaphysis
   c. Epiphyseal plate
   d. Lamella

_____ 18. What is the term for a hole or opening in a bone for blood vessels and nerves?
   a. Fossa
   b. Suture
   c. Process
   d. Foramen

_____ 19. The cartilage that attaches the ribs to the sternum is called the
   a. Articular cartilage
   b. Intervertebral disk
   c. Epiphyseal plate
   d. Costal cartilage

_____ 20. Which of the following bones forms the back of the skull?
   a. Occipital bone
   b. Parietal bone
   c. Temporal bone
   d. Ethmoid bone

_____ 21. Which of the following bones forms the posterior aspect of the hard palate?
   a. Maxilla
   b. Vomer
   c. Palatine
   d. Sphenoid

_____ 22. The rib cage is composed of
   a. 24 ribs and the scapula
   b. 12 ribs and the clavicle
   c. 12 pairs of ribs and the scapula
   d. 12 pairs of ribs and the sternum

_____ 23. What is the term for the membrane around the shaft of a long bone?
   a. Endosteum
   b. Periosteum
   c. Osteoclast
   d. Osteocyte

_____ 24. Which bone is the largest of the tarsal bones?
   a. Malleolus
   b. Acetabulum
   c. Calcaneus
   d. Metatarsal

_____ 25. The coxal bones are more commonly known as the
   a. Sternal bones
   b. Hand bones
   c. Hip bones
   d. Wrist bones

FILL IN THE BLANKS

In the space provided, write the word or phrase that best completes each sentence. Not all words or phrases are used.

26. _____ are immature bone cells that create new bone.

27. _____ are cells that digest old bone in preparation for new bone growth.

28. _____ are mature bone cells.

29. _____ is the process of blood formation.

30. The fibrous bands that attach bone to bone are _____.

31. The _____ is commonly known as the shaft of a long bone.

32. The rounded end of the long bone is called the _____.

33. _____ bone is where red bone marrow is located.

34. Yellow bone marrow is located in the _____.

35. The _____ is the tough fibrous membrane that covers the shaft of the long bone.

a. Calcification
b. Cancellous
c. Diaphysis
d. Epiphysis
e. Hematopoiesis
f. Leukemia
g. Ligaments
h. Medullary cavity
i. Metaphysis
j. Osteoblasts
k. Osteoclasts
l. Osteocytes
m. Periosteum
n. Spongy
o. Tendons

SHORT ANSWER

Write the answer to each statement on the lines provided.

36. Differentiate the types of bone shapes.

_____

_____

_____

_____

37. Describe the bones of the axial skeleton.

_____

_____

_____

_____

38. Describe the bones of the pectoral girdle.

_____

_____

_____

_____

39. Describe the articulation between the upper and lower leg.

_____

_____

_____

_____

LABELING

Follow the directions and write the answers on the lines provided.

40. The following figure shows the structure of a long bone. Identify these terms in the figure by writing them on the lines provided: *articular cartilage, compact bone, diaphysis, distal epiphysis, epiphyseal plates, medullary cavity, periosteum, proximal epiphysis, space occupied by red marrow, spongy bone, yellow marrow.*

    a. _____

    b. _____

    c. _____

64   UNIT 2   Concepts of Common Illness by System

d. _____

e. _____

f. _____

g. _____

h. _____

i. _____

j. _____

k. _____

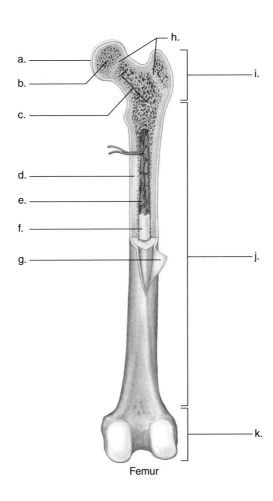

Femur

41. The following figure shows anterior and posterior views of the human skeleton. Identify these terms in the figure by writing them on the lines provided: *carpals, clavicle, coccyx, coxa, femur, fibula, forearm, humerus, hyoid bone, metacarpals, metatarsals, patella, phalanges, radius, ribs, sacrum, scapula, sternum, tarsals, tibia, vertebral column, ulna.* (A word may be used more than once or not at all.)

a. _____

b. _____

c. _____

d. _____

e. _____

f. _____

g. _____

h. _____

i. _____

j. _____

k. _____

l. _____

m. _____

n. _____

o. _____

p. _____

q. _____

r. _____

s. _____

t. _____

u. _____

v. _____

w. _____

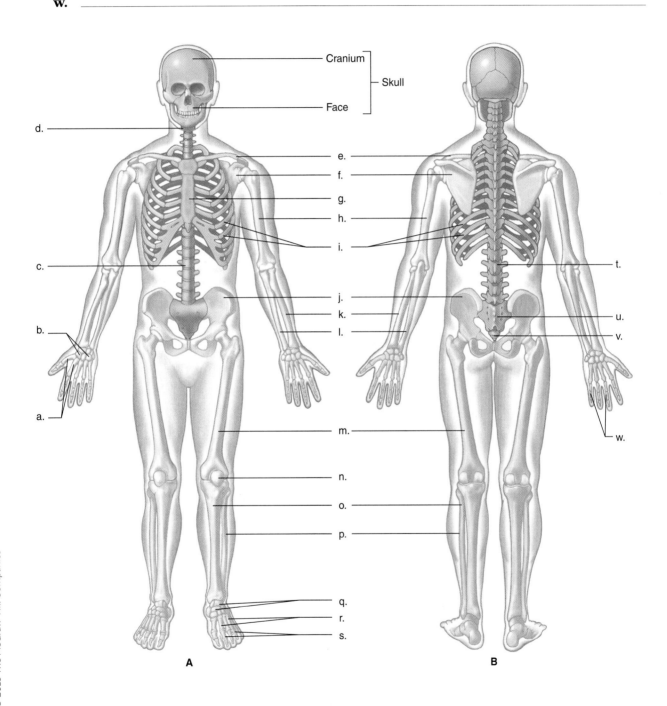

CHAPTER 8   The Skeletal System   67

42. The following figure shows a lateral view of the human skull. Identify these terms in the figure by writing them on the lines provided: *external auditory meatus, frontal bone, lacrimal bone, mandible, mastoid process, maxilla, nasal bone, occipital bone, parietal bone, styloid process, temporal bone, zygomatic bone.* (A word may be used more than once or not at all.)

a. _____

b. _____

c. _____

d. _____

e. _____

f. _____

g. _____

h. _____

i. _____

j. _____

k. _____

l. _____

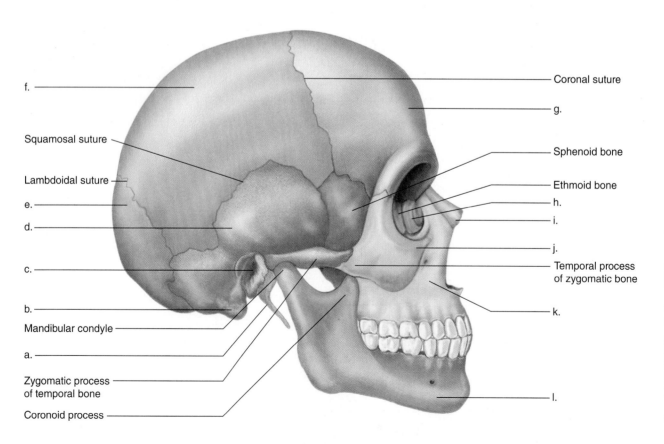

68 UNIT 2 Concepts of Common Illness by System

critical thinking / application

CRITICAL THINKING

Write the answer to each question or statement on the lines provided.

43. What roles do osteoblasts and osteoclasts play throughout life?

44. What is the function of vitamin D in bone health and growth?

45. Explain the importance of the pectoral and pelvic girdles. What bones are included in each?

46. Describe the three common vertebral deformities and give at least one causative factor for each.

APPLICATION

Follow the directions for the application. Write the answer to each statement on the lines provided.

47. Inflammation is an "enemy" of bones and joints. Explain how inflammation affects bones and joints in the following inflammatory diseases.

   a. Bursitis _____

   b. Osteoarthritis _____

   c. Rheumatoid arthritis _____

   d. Gouty arthritis _____

case studies

Write your response to each case study question on the lines provided.

48. A 25-year-old woman is concerned that she may have osteoporosis because her mother has the disorder. Is the 25-year-old likely to have osteoporosis? What would you tell her about preventing the disease?

49. A young woman is pregnant and neglects to make sure that her diet has sufficient amounts of calcium. How is her diet likely to affect her bones and the bones of her fetus?

50. A 60-year-old man has just been diagnosed with gouty arthritis. What can you tell him about his condition and preventive measures he may take to prevent attacks?

pathophysiology

Follow the instructions for the statement.

51. Fill in the missing cells.

Disease	Etiology	Signs and Symptoms	Treatment
Osteoarthritis			
Gout			
Carpal tunnel			
Scoliosis			
Osteosarcoma			

The Muscular System 9

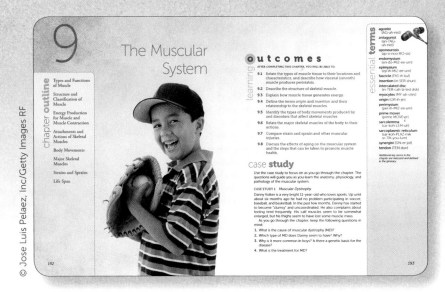

vocabulary review

MATCHING

Match the key terms in the right column with the definitions in the left column by writing the letter of the correct answer in the space provided.

_____ 1. Helper muscle in skeletal muscle movement

_____ 2. A grouping of muscle fibers

_____ 3. A condition produced by buildup of lactic acid in muscle

_____ 4. A structure that connects groups of cardiac muscle

_____ 5. Tough connective tissue that attaches muscle to bone

_____ 6. Connective tissue that surrounds the entire muscle

_____ 7. The cell membrane of a muscle fiber

_____ 8. The muscle primarily responsible for skeletal muscle movement

_____ 9. An energy-producing process for muscles that requires oxygen

_____ 10. A muscle that works in opposition to a prime mover

a. Aerobic respiration
b. Antagonist
c. Epimysium
d. Fascicle
e. Intercalated disc
f. Muscle fatigue
g. Prime mover
h. Sarcolemma
i. Synergist
j. Tendon

content review

MULTIPLE CHOICE

In the space provided, write the letter of the choice that best completes each statement or answers each question.

_____ 11. Which of the following is *not* a function of muscle?
   a. Production of body movement
   b. Stabilization of joints
   c. Control of body openings
   d. All of the above are functions of muscle

_____ 12. Which of the following is true about visceral muscle?
   a. It is voluntary.
   b. It is striated.
   c. It is slow to contract and relax.
   d. It is attached to bones.

_____ 13. Which of the following is a connective tissue that surrounds fascicles?
   a. Fascia
   b. Epimysium
   c. Endomysium
   d. Perimysium

_____ 14. Which disease is commonly called lockjaw?
   a. Botulism
   b. Tetanus
   c. Trichinosis
   d. Encephalitis

_____ 15. Torticollis is also known as
   a. Wry neck
   b. Encephalitis
   c. Cervicalgia
   d. Angina

_____ 16. Which of the following is the action of pointing the toes upward?
   a. Inversion
   b. Eversion
   c. Plantar flexion
   d. Dorsiflexion

_____ 17. Which of the following is the action of moving a body part posteriorly?
   a. Elevation
   b. Retraction
   c. Protraction
   d. Depression

_____ 18. Which muscle closes the jaw?
   a. Platysma
   b. Masseter
   c. Sternocleidomastoid
   d. Orbicularis oris

_____ 19. Which of the following structures connect cardiac muscle fibers to each other?
   a. Tendons
   b. Ligaments
   c. Intracalated discs
   d. Intercalated discs

_____ 20. Which term is another name for a heart muscle cell?
   a. Cardiomyocyte
   b. Sarcolemma
   c. Sarcoplasm
   d. Myofibril

_____ 21. Peristalsis describes the contraction of which of the following muscle types?
   a. Skeletal
   b. Visceral (smooth)
   c. Multiunit (smooth)
   d. Cardiac

_____ 22. Which term refers to the loss of a muscle's ability to contract?
   a. Oxygen debt
   b. Lactic acid production
   c. Muscle fatigue
   d. Aerobic respiration

FILL IN THE BLANKS

In the space provided, write the word or phrase that best completes each sentence. Not all words or phrases are used. Words may be used more than once.

23. _____ is the pinkish pigment of muscle that stores oxygen.

24. The attachment of muscle to the more movable bone is known as the _____.

25. The neurotransmitter _____ causes a skeletal muscle response.

26. The rhythmic contraction that moves substances through tubes of the body is called _____.

27. The site of attachment of a muscle to the less movable bone during a muscle contraction is called the _____.

28. _____ is the enzyme responsible for skeletal muscle relaxation.

29. The _____ muscle attaches to the temporal bone and the coronoid process of the mandible and elevates and retracts the mandible.

30. The _____ describes the aerobic process of creating ATP for muscle energy.

31. The byproduct of pyruvic acid conversion is _____, which occurs when muscle is low on oxygen.

32. Dense, irregular connective tissue surrounding an entire muscle is known as _____.

33. A(n) _____ is a broad sheetlike structure made of connective tissue and typically attaches muscles to other muscles.

34. A(n) _____ assists a prime mover in the muscle movement.

35. _____ is released by skeletal muscles to break down acetylcholine after muscle contraction has occurred.

a. Acetylcholine
b. Acetylcholinesterase
c. Agonist
d. Aponeurosis
e. Carotene
f. Epimysium
g. Glycolysis
h. Insertion
i. Krebs cycle
j. Lactic acid
k. Lipase
l. Myoglobin
m. Origin
n. Peristalsis
o. Segmentation
p. Synergist
q. Temporalis

SHORT ANSWER

Write the answer to each statement on the lines provided.

36. Explain a neuromuscular junction.

37. Describe the differences between smooth muscle, skeletal muscle, and cardiac muscle.

38. Explain the difference between the insertion of a muscle and the origin of a muscle.

39. Compare flexion and extension of the leg with flexion and extension of the forearm.

40. Describe how energy is obtained by muscles for contraction.

LABELING

Follow the directions and write the answers on the lines provided.

41. The following figure shows the muscular system. Identify the following terms on the lines provided: *biceps brachii, deltoid, external oblique, gastrocnemius, gluteus maximus, gluteus medius, latissimus dorsi, pectoralis major, rectus abdominis, rectus femoris, semitendinosus, trapezius, triceps brachii, vastus lateralis.*

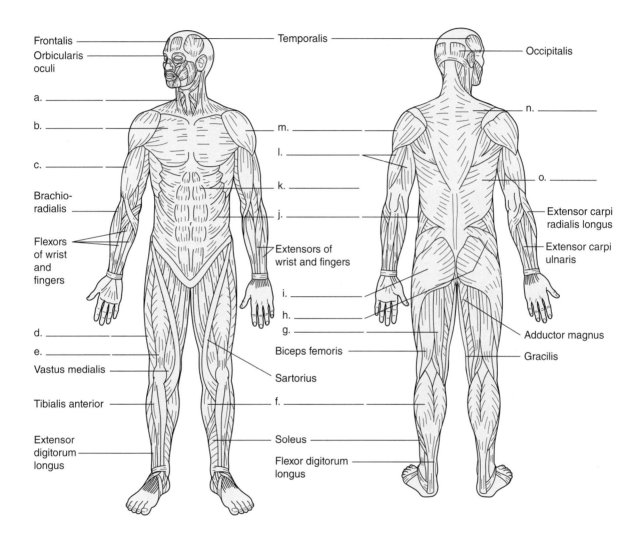

a. _____

b. _____

c. _____

d. _____

e. _____

f. _____

CHAPTER 9   The Muscular System   77

g. _____

h. _____

i. _____

j. _____

k. _____

l. _____

m. _____

n. _____

o. _____

critical thinking / application

CRITICAL THINKING

Write the answer to each question on the lines provided.

42. Norepinephrine works with the sympathetic nervous system. How does its presence in the bloodstream affect the heart?

_____

_____

_____

_____

43. When a person is exercising, his or her breathing rate increases to ensure the adequate delivery of oxygen to skeletal muscle tissues. Why does the breathing rate stay increased for a period of time even after a person stops exercising?

_____

_____

_____

_____

44. When a muscle is overworked, it often becomes sore and cramps. Why does massaging a muscle help reduce muscle soreness?

_____
_____
_____
_____

45. In cooler weather, mothers often tell children to put on a sweater or "get up and move around" to warm up. Why do they think movement will warm up their children?

_____
_____
_____
_____

46. Why do cardiac muscle fibers have intercalated discs?

_____
_____
_____
_____

APPLICATION

Follow the directions for the application. Write the answer to each statement on the lines provided.

47. For each of the following skeletal muscle actions, list the prime mover, at least one synergist, and the antagonist.

    a. Bending the leg at the knee _____

_____
_____

b. Raising the arm laterally at the shoulder _____

c. Bending the hand at the wrist _____

case studies

Write your response to each case study question on the lines provided.

48. A friend calls and tells you that she thinks she has a muscle strain. You tell her that immediate RICE treatment is recommended. What does this mean?

49. A friend has a child that has just been diagnosed with muscular dystrophy. What do you tell her it is, the cause, and possible treatments?

50. A young man is hospitalized for seizures after running his first marathon. He is experiencing muscle fatigue and cramps. What is the difference between the two?

_____

_____

_____

_____

pathophysiology

Follow the instructions for the statement.

51. Fill in the missing cells.

Disease	Etiology	Signs and Symptoms	Treatment
Fibromyalgia			
Myasthenia gravis			
Sprain			

Blood and Circulation

10

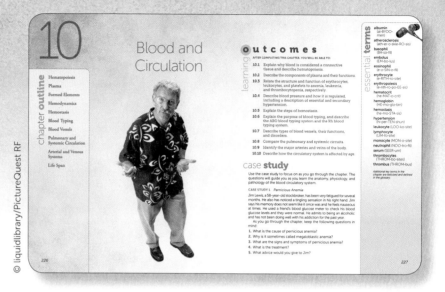

vocabulary review

MATCHING

Match the key terms in the right column with the definitions in the left column by writing the letter of the correct answer in the space provided.

_____ 1. A mature red blood cell

_____ 2. A cell fragment involved in hemostasis

_____ 3. Smallest, most abundant of the plasma proteins

_____ 4. Blood plasma minus its clotting proteins

_____ 5. Iron-containing pigment in red blood cells responsible for transporting oxygen

_____ 6. A type of white blood cell involved in hypersensitivity reactions

_____ 7. A significantly elevated WBC count

_____ 8. Elevated blood pressure above normal

_____ 9. Stoppage of blood flow by natural or artificial means

_____ 10. An accumulation of cholesterol and smooth muscle fibers in the tunica media of an artery

_____ 11. A white blood cell that transforms into a macrophage when it leaves the circulation

_____ 12. The percentage of blood made up of red blood cells

_____ 13. A type of white blood cell that has phagocytic capabilities and with a multilobed nucleus

_____ 14. Condition caused by an accumulation of bilirubin in the blood

_____ 15. A type of white blood cell involved in allergic conditions and parasitic infections

a. Albumin
b. Atherosclerosis
c. Basophil
d. Eosinophil
e. Erythrocyte
f. Hematocrit
g. Hemoglobin
h. Hemostasis
i. Hypertension
j. Jaundice
k. Leukocytosis
l. Monocyte
m. Neutrophil
n. Serum
o. Thrombocyte

content review

MULTIPLE CHOICE

In the space provided, write the letter of the choice that best completes each statement or answers each question.

_____ 16. The formation of red and white blood cells is called
   a. Anemia
   b. Erythropoiesis
   c. Hematopoiesis
   d. Leukopenia

_____ 17. During fetal development, red blood cells are made in all of the following *except* the
   a. Spinal cord
   b. Yolk sac
   c. Liver
   d. Spleen

_____ 18. What is the average life span of a red blood cell?
   a. 180 days
   b. 90 days
   c. 150 days
   d. 120 days

_____ 19. What is the term for the formation of red blood cells?
   a. Leukocytosis
   b. Erythropoiesis
   c. Leukemia
   d. Anemia

_____ 20. Which of the following is the most abundant plasma protein?
   a. Collagen
   b. Albumin
   c. Hemoglobin
   d. Fibrinogen

_____ 21. Lipids must combine with molecules called _____ to be transported in the blood.
   a. Lipoproteins
   b. Triglycerides
   c. Macroglobulins
   d. Micelles

_____ 22. Gases dissolved in plasma include all of the following *except*
   a. Nitrogen
   b. Oxygen
   c. Helium
   d. Carbon dioxide

_____ 23. Which of the following is the blood test to measure cholesterol?
   a. WBC
   b. Lipoprotein profile
   c. Blood cell index
   d. Fat index

_____ 24. The iron-containing pigment in erythrocytes responsible for transporting most of the oxygen in the blood is called
   a. Ferric nitrate
   b. Hemoglobin
   c. Iron phosphate
   d. Hematocrit

UNIT 2 Concepts of Common Illness by System

_____ **25.** Which of the following substances emulsifies or breaks apart fat molecules?
   a. Biliverdin
   b. Iron
   c. Hemoglobin
   d. Bile

_____ **26.** When the blood has less than its normal oxygen-carrying capacity, what occurs?
   a. Leukocytosis
   b. Leukemia
   c. Anemia
   d. Erythrocytosis

_____ **27.** When the bone marrow is destroyed, the result is _____ anemia.
   a. Aplastic
   b. Iron deficiency
   c. Leukoplastic
   d. Hemolytic

_____ **28.** A deficiency of folic acid causes what type of anemia?
   a. Microcytic
   b. Iron deficiency
   c. Megaloblastic
   d. Microblastic

_____ **29.** All of the following white blood cells are granulocytes *except*
   a. Basophils
   b. Eosinophils
   c. Neutrophils
   d. Lymphocytes

_____ **30.** Which are the least numerous white blood cells in the blood under normal conditions?
   a. Erythrocytes
   b. Basophils
   c. Eosinophils
   d. Neutrophils

_____ **31.** _____ is a neoplastic condition in which the bone marrow produces a large number of white blood cells that are not normal.
   a. Anemia
   b. Leukocytosis
   c. Erythrocytosis
   d. Leukemia

_____ **32.** Which of the following is a cell fragment involved in blood clotting?
   a. Fibrinogen
   b. Reticulocyte
   c. Platelet
   d. Fibrin

_____ **33.** _____ is a condition where there are too few thrombocytes causing abnormal bleeding.
   a. Thrombocytopenia
   b. Thrombocytosis
   c. Anemia
   d. Leukemia

_____ **34.** Hemophilia is a group of blood disorders that are classified as
   a. X-linked
   b. Y-linked
   c. Autosomal dominant
   d. Autosomal recessive

_____ **35.** Which of the following is defined as a disease or condition of unknown cause?
   a. Malignant
   b. Idiopathic
   c. Benign
   d. Nosocomial

FILL IN THE BLANKS

In the space provided, write the word or phrase that best completes each sentence. Not all words or phrases are used. Words may be used more than once.

36. _____ may be used to help control blood hypertension by decreasing blood volume by increasing urination.

37. _____ is the stoppage of blood flow by natural or artificial means.

38. The formation of a blood clot is called blood _____.

39. _____ is the mesh web that entraps platelets and blood cells to form a blood clot.

40. A(n) _____ is anything such as a blood clot, air bubble, fat, or other foreign substance that is transported by the blood.

41. Hypoperfusion to tissues resulting in decreased oxygen and nutrient delivery can result in _____.

42. _____ is the clumping of RBCs that can occur if an individual is given a blood type during a transfusion that is different than his or her own.

43. People with type AB blood are called _____ because they can receive type A, B, AB, or O blood types.

44. A person with type A blood has the _____ antigen.

45. A person with type B blood has the _____ antibody.

a. A
b. Agglutination
c. Anti-A
d. B
e. Coagulation
f. Diuretics
g. Embolus
h. Fibrin
i. Hematopoiesis
j. Hemostasis
k. Shock
l. Thrombus
m. Universal donor
n. Universal recipients

SHORT ANSWER

Write the answer to each statement on the lines provided.

46. Describe the components of plasma and their functions.

_____
_____
_____
_____

47. Explain the difference between essential and secondary hypertension.

48. Explain the purpose of blood typing and describe the ABO blood typing system.

49. Compare the pulmonary and systemic circulations.

50. Describe the differences between arteries and veins.

LABELING

Follow the directions and write the answers on the lines provided.

51. In the following centrifuged blood sample, identify these terms by writing them on the lines provided: *buffy coat, plasma, red blood cells.*

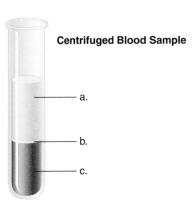

a. _____

b. _____

c. _____

52. For the following figures showing white blood cells, identify these terms by writing them on the lines provided: *basophil, eosinophil, lymphocyte, monocyte, neutrophil.*

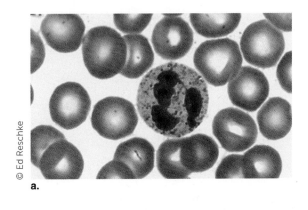

a.

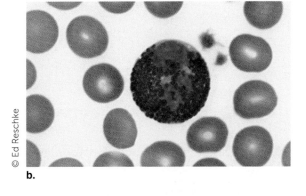

b.

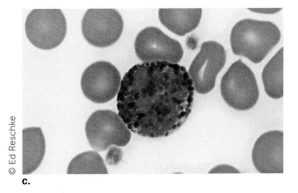

c.

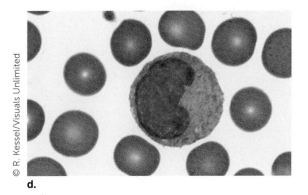

d.

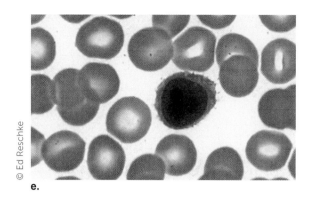

e.

a. _____

b. _____

c. _____

d. _____

e. _____

53. For the following figure showing pulse sites, identify these terms by writing them on the lines provided: *brachial artery, carotid artery, dorsal pedis artery, facial artery, femoral artery, popliteal artery, posterior tibial artery, radial artery, temporal artery.*

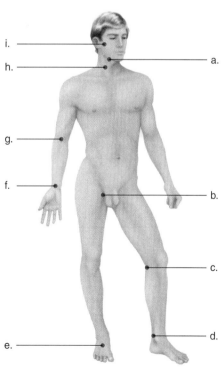

From *Hole's Human Anatomy & Physiology,* 12e, by Shier/Butler/and Lewis. Copyright © 2009. Reprinted by permission of McGraw-Hill Companies Inc.

a. _____

b. _____

c. _____

d. _____

e. _____

f. _____

g. _____

h. _____

i. _____

critical **thinking** / **application**

CRITICAL THINKING

Write the answer to each statement on the lines provided.

54. Compare sickle cell trait and sickle cell disease.

_____

_____

_____

_____

APPLICATION

Follow the directions for the application. Write the answer to the question on the lines provided.

55. What is the treatment for megaloblastic anemia?

_____

_____

_____

_____

case studies

Write your response to each case study question on the lines provided.

56. A young child of Mediterranean descent has been suffering headaches and nausea and loss of appetite. There also seems to be some bronzing of the child's skin. Based on the symptoms and the child's ethnicity, describe the condition that she may be suffering from.

_____

_____

_____

_____

57. A teenager suffers from allergies. Describe what you would expect to see on a WBC count if taken during an allergy attack.

_____

_____

_____

_____

pathophysiology

Follow the instructions for the statement.

58. Fill in the missing cells.

Disease	Etiology	Signs and Symptoms	Treatment
Abdominal aortic aneurysm (AAA)			
Varicose veins			
Megaloblastic anemia			

CHAPTER 10 Blood and Circulation

The Cardiovascular System

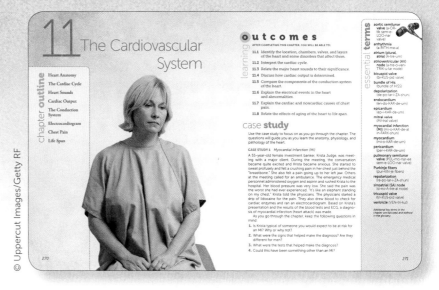

vocabulary review

MATCHING

Match the key terms in the right column with the definitions in the left column by writing the letter of the correct answer in the space provided.

_____ 1. A change in the resting potential of a cell that results in an action potential

_____ 2. The inner layer of the heart that also covers the heart valves

_____ 3. The primary pacemaker of the heart

_____ 4. The secondary pacemaker of the heart

_____ 5. A membrane that surrounds the heart

_____ 6. Blood vessels that carry oxygenated blood from the lungs to the left atrium

_____ 7. The large artery that comes off the left ventricle

_____ 8. The anatomical space between the pleurae of the lungs that contains several structures including the heart

_____ 9. An irregular heart rhythm

_____ 10. The thin outer layer of the heart wall

_____ 11. Heart valve located between the left atrium and the left ventricle

_____ 12. Part of the electrical conduction system of the heart; extends down the interventricular septum

_____ 13. Atrioventricular valve on the right side of the heart

_____ 14. An inferior chamber of the heart

_____ 15. A recording of the electrical changes that accompany the cardiac cycle

a. Aorta
b. Arrhythmia
c. AV node
d. Bundle of His
e. Depolarization
f. Electrocardiogram
g. Endocardium
h. Epicardium
i. SA node
j. Mitral valve
k. Pericardium
l. Pulmonary veins
m. Thoracic mediastinum
n. Tricuspid valve
o. Ventricles

content review

MULTIPLE CHOICE

In the space provided, write the letter of the choice that best completes each statement or answers each question.

_____ 16. In an ECG, what does the QRS complex represent?
   a. Atrial depolarization
   b. Ventricular depolarization
   c. Atrial repolarization
   d. Ventricular repolarization

_____ 17. Which of the following is the wall between the pumping chambers of the heart?
   a. Interventricular septum
   b. Intraventricular septum
   c. Interatrial septum
   d. Atrioventricular septum

_____ 18. What does the P wave on an ECG reading represent?
   a. Atrial depolarization
   b. Atrial repolarization
   c. Ventricular depolarization
   d. Ventricular repolarization

_____ 19. What is the term for the space between the right and left lungs?
   a. Dorsal cavity
   b. Thoracic mediastinum
   c. Abdominal cavity
   d. Phrenic cavity

_____ 20. Which is the most muscular heart chamber?
   a. Right atrium
   b. Left atrium
   c. Right ventricle
   d. Left ventricle

_____ 21. Which chamber of the heart forms the anterior surface of the heart?
   a. Right atrium
   b. Left atrium
   c. Right ventricle
   d. Left ventricle

_____ 22. How many valves do the pulmonary veins have?
   a. None
   b. One
   c. Two
   d. Four

_____ 23. How many valves are in the heart?
   a. None
   b. One
   c. Two
   d. Four

_____ 24. How many cusps does the mitral valve have?
   a. One
   b. Two
   c. Three
   d. Four

_____ 25. What is the outermost layer of the heart called?
   a. Endocardium
   b. Myocardium
   c. Epicardium
   d. Neurocardium

FILL IN THE BLANKS

In the space provided, write the word or phrase that best completes each sentence. Not all words or phrases are used.

26. The _____ is the muscular layer of the heart.

27. Blood returns to the _____ from the lungs.

28. The _____ is the primary pacemaker of the heart.

29. _____ carry blood away from the heart.

30. _____ fluid is found within the pericardial sac.

31. Parasympathetic innervation of the heart is provided by the _____ nerve.

32. _____ is another name for a heart attack.

33. An increase in body temperature will _____ heart rate.

34. Heart sounds are heard through _____ of the heart with a stethoscope.

35. _____ is the volume of blood pumped out of the left or right ventricle each minute.

a. Arteries
b. Atrioventricular node
c. Auscultation
d. Cardiac output
e. Endocardium
f. Facial
g. Increase
h. Left atrium
i. Myocardial infarction
j. Myocardium
k. Right atrium
l. Serous
m. Sinoatrial node
n. Stroke volume
o. Vagus
p. Veins

SHORT ANSWER

Write the answer to each statement on the lines provided.

36. Differentiate between the major heart sounds and their relation to the cardiac cycle.

_____
_____
_____
_____

37. Explain why ventricular fibrillation may cause sudden cardiac death.

38. Explain cardiac output and how it is calculated.

39. Describe the components of the electrical conduction system of the heart.

40. Explain the components of an ECG reading.

LABELING

Follow the directions and write the answers on the lines provided.

41. In the following illustration showing the structure of the heart wall, identify these terms by writing them on the lines provided: *coronary blood vessel, endocardium, epicardium, fibrous pericardium, myocardium, parietal pericardium, pericardial cavity.*

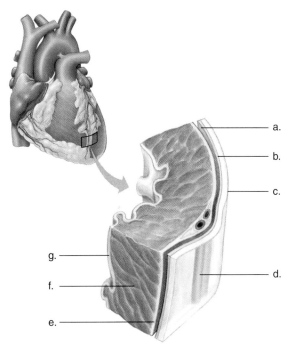

From *Hole's Human Anatomy & Physiology*, 12e, by Shier/Butler/and Lewis. Copyright © 2009. Reprinted by permission of McGraw-Hill Companies Inc.

a. _____

b. _____

c. _____

d. _____

e. _____

f. _____

g. _____

42. In the following illustration of the heart, identify these terms by writing them on the lines provided: *aorta, bicuspid valve, chordae tendinae, inferior vena cava, interventricular septum, left atrium, left pulmonary artery, left pulmonary veins, left ventricle, papillary muscles, pulmonary trunk, pulmonary valve, right atrium, right pulmonary artery, right pulmonary veins, right ventricle, superior vena cava, tricuspid valve.*

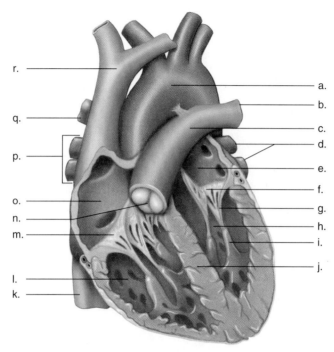

From *Hole's Human Anatomy & Physiology*, 12e, by Shier/Butler/and Lewis.
Copyright © 2009. Reprinted by permission of McGraw-Hill Companies Inc.

a. _____

b. _____

c. _____

d. _____

e. _____

f. _____

g. _____

h. _____

i. _____

98 UNIT 2 Concepts of Common Illness by System

j. _____

k. _____

l. _____

m. _____

n. _____

o. _____

p. _____

q. _____

r. _____

43. In the following illustration showing the flow of impulses through the cardiac conduction system, identify these terms by writing them on the lines provided: *AV bundle, AV node, interatrial septum, interventricular septum, junctional fibers, left bundle branch, Purkinje fibers, right bundle branch, SA node.*

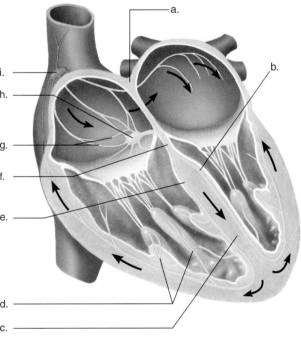

From *Hole's Human Anatomy & Physiology,* 12e, by Shier/Butler/and Lewis.
Copyright © 2009. Reprinted by permission of McGraw-Hill Companies Inc.

a. _____

b. _____

c. _____

d. _____

e. _____

f. _____

g. _____

h. _____

i. _____

critical **thinking / application**

CRITICAL THINKING

Write the answer to each question or statement on the lines provided.

44. The left ventricular wall is more muscular than the right ventricular wall. Why do you think this is necessary?

_____

_____

_____

_____

45. Why is it important for tricuspid and bicuspid valves to close completely when the ventricles contract?

_____

_____

_____

_____

46. Patients with angina pectoris are given sublingual nitroglycerin. Why is this medication helpful to them?

47. Explain the effects of aging on the cardiovascular system.

APPLICATION

Write the answer to the statement on the lines provided.

48. Trace the flow of blood from the superior and inferior vena cava through the heart and lungs and out the aorta. Do not forget to mention the valves.

case studies

Write your response to each case study question on the lines provided.

49. The right side of the heart in a 68-year-old patient is failing. What are the effects on various organs in the body if the condition is not treated?

_____

_____

_____

_____

50. A 56-year-old woman thinks she is having a heart attack. She has just eaten a spicy Italian meal and her chest pains seem to get worse when she bends forward or lies down. What would you tell her?

_____

_____

_____

_____

pathophysiology

Follow the instructions for the statement.

51. Fill in the missing cells.

Disease	Etiology	Signs and Symptoms	Treatment
Myocardial infarction			
Angina pectoris			
Congestive heart failure			
Endocarditis			

102   UNIT 2   Concepts of Common Illness by System

The Lymphatic and Immune Systems

12

vocabulary review

MATCHING

Match the key terms in the right column with the definitions in the left column by writing the letter of the correct answer in the space provided

_____ 1. Soft bilobed organ located just below the thyroid gland in the mediastinum
_____ 2. A disease-causing agent
_____ 3. A foreign substance in the body capable of eliciting an immune response
_____ 4. A cell that produces antibodies
_____ 5. Antibodies produced by plasma cells
_____ 6. An excessive immune response to a stimulus
_____ 7. A life-threatening allergic reaction
_____ 8. A failure of self-recognition by the body where it attacks its own antigens
_____ 9. Profuse sweating
_____ 10. Substances that can trigger an allergic response
_____ 11. Condition caused by blockage of the lymphatic vessels
_____ 12. Lymphoid tissue located in the oral and nasal cavities
_____ 13. Presence of a pathogen in or on the body
_____ 14. A type of T cell that protects the body against viruses and cancer cells
_____ 15. A hormone that can cause vasoconstriction
_____ 16. Presence of a fever
_____ 17. Largest lymphatic vessel in the body
_____ 18. The response that occurs the first time a person is exposed to an antigen
_____ 19. The largest lymphatic organ
_____ 20. Another name for the pharyngeal tonsils

a. Adenoids
b. Allergen
c. Allergic reaction
d. Anaphylaxis
e. Antigen
f. Autoimmune disease
g. Cytotoxic T cells
h. Diaphoresis
i. Epinephrine
j. Febrile
k. Immunoglobulins
l. Infection
m. Lymphedema
n. Pathogen
o. Plasma cell
p. Primary immune response
q. Spleen
r. Thoracic duct
s. Thymus
t. Tonsils

content review

MULTIPLE CHOICE

In the space provided, write the letter of the choice that best completes each statement or answers each question.

_____ 21. A buildup of fluids in the interstitial spaces is called
   a. Plasma
   b. Lymph
   c. Edema
   d. Serum

_____ 22. Lymph from the body eventually drains into the
   a. Left atrium
   b. Right atrium
   c. Right ventricle
   d. Left ventricle

_____ 23. Nerves and blood vessels enter a lymph node at the
   a. Hilum
   b. Medulla
   c. Cortex
   d. Metaphysis

_____ 24. What is the term for cells derived from B lymphocytes that produce antibodies?
   a. Basophils
   b. Monocytes
   c. Eosinophils
   d. Plasma cells

_____ 25. Which of the following trigger a stronger immune response the next time the person is exposed to the same antigen?
   a. Memory cells
   b. Erythrocytes
   c. Basophils
   d. Mast cells

_____ 26. Natural killer cells secrete _____ that can punch holes in the membranes of harmful cells, causing the cells to lyse or rupture.
   a. Immunoglobulins
   b. Perforins
   c. Lysozymes
   d. Antibodies

_____ 27. Which of the following is *not* an immunoglobulin?
   a. IgA
   b. IgI
   c. IgM
   d. IgE

_____ 28. Which antibody protects against parasite infections?
   a. IgM
   b. IgA
   c. IgE
   d. IgD

_____ 29. Who developed a vaccine against smallpox?
   a. Edward Jenner
   b. Edward Hughes
   c. John Edwards
   d. Martha Edwards

_____ 30. Which substances can trigger allergic responses?
   a. Antibodies
   b. Toxoids
   c. Allergens
   d. Hemoglobins

31. In systemic lupus erythematosis (SLE), which antibodies attack the cells' DNA?
    a. Rh
    b. Anti-nuclear
    c. ABO
    d. Cytoplasmic

32. Which of the following is the most common symptomatic finding in SLE?
    a. Butterfly rash
    b. Arthritis
    c. Headaches
    d. Shortness of breath

33. Which of the following is a rare type of cancer that is more common in HIV-infected individuals?
    a. Squamous cell carcinoma
    b. Leukemia
    c. Rhabdomyoma
    d. Kaposi sarcoma

34. All of the following are true of acquired immunity *except*
    a. It is slower to respond than innate immunity
    b. It is also called specific immunity
    c. We are born with it
    d. It remembers dangerous microbes that it has encountered in the past

35. The blockage of lymphatic vessels is referred to as
    a. Lymphedema
    b. Thrombosis
    c. Embolism
    d. DVT

36. Lymph flow is
    a. Unidirectional toward the heart
    b. Unidirectional away from the heart
    c. Bidirectional flowing mostly toward the heart
    d. Bidirectional flowing mostly away from the heart

37. What is the outer portion of the thymus called?
    a. Capsule
    b. Cortex
    c. Medulla
    d. Hilum

38. All of the following are tonsils *except*
    a. Lingual tonsils
    b. Pharyngeal tonsils
    c. Laryngeal tonsils
    d. Palatine tonsils

39. The presence of a pathogen in or on the body is called what?
    a. Invasion
    b. Infection
    c. Involution
    d. Immigration

40. What is the term for a disease that is transmissible from an animal to a human?
    a. Marginal
    b. Biological
    c. Pathological
    d. Zoonotic

FILL IN THE BLANKS

In the space provided, write the word or phrase that best completes each sentence. Not all words or phrases are used.

41. The _____, which is part of the lymphatic system, is found in the left upper quadrant of the abdominal cavity.

42. After adolescence, the thymus starts to _____ or involute.

43. _____ are lymph nodules located in the small intestine.

44. _____ in tears destroy pathogens that come in contact with the eye surface.

45. Lysozymes and interferon are examples of _____ barriers to prevent infection.

46. Inflammation is a(n) _____ response.

47. Both B cells and T cells are produced in the red _____.

48. Every human being except identical twins has a unique _____ that acts as a "cell identity marker."

49. _____ is found in breast milk, sweat, tears, and saliva and protects the mucosa against bacteria and viruses.

50. An injection of _____ is the treatment of choice for anaphylaxis.

a. Atrophy
b. Bone marrow
c. Cellular
d. Chemical
e. Enzyme
f. Epinephrine
g. Hypertrophy
h. IgA
i. IgE
j. Lysozyme
k. Major histo-compatibility complex
l. Norepinephrine
m. Peyer's patches
n. Spleen
o. Thymus
p. Vascular

SHORT ANSWER

Write the answer to each statement on the lines provided.

51. Compare and contrast interstitial fluid and lymph.

_____

_____

_____

_____

52. Explain the different types of grafts or transplants.

53. Summarize the structure and function of the thymus and spleen.

54. Differentiate between specific and nonspecific immunity.

55. Describe the structure and function of lymph nodes.

LABELING

Follow the directions and write the answers on the lines provided.

56. Using the following figure, identify these terms by writing them on the lines provided: *axillary lymph nodes, cisterna chyli, inguinal lymph nodes, left subclavian vein, thoracic duct.*

a. _____

b. _____

c. _____

d. _____

e. _____

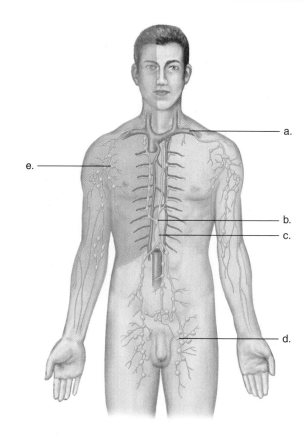

57. For the following figure, identify the structures using these terms: *axillary lymph nodes, lymphatics of mammary gland, right internal jugular vein, right lymphatic duct, right subclavian vein.*

a. _____

b. _____

c. _____

d. _____

e. _____

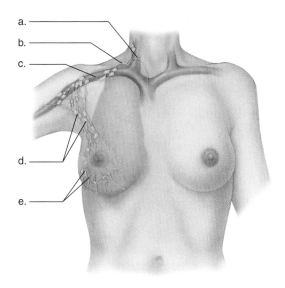

From *Hole's Human Anatomy & Physiology*, 12e, by Shier/Butler/and Lewis. Copyright © 2009. Reprinted by permission of McGraw-Hill Companies Inc.

critical **thinking / application**

CRITICAL THINKING

Write the answer to the question on the lines provided.

58. Why is someone with HIV/AIDS more susceptible to opportunistic infections?

_____

_____

_____

_____

APPLICATION

Write the answer to the question on the lines provided.

59. If HIV can be transmitted through breast milk, what would be the recommendations for a poor woman living in Africa who is HIV-positive and currently breast-feeding her infant?

_____

_____

_____

_____

CHAPTER 12   The Lymphatic and Immune Systems

case studies

Write your response to each case study question on the lines provided.

60. A young woman has just given birth to her first child. How would you explain the benefits of breast-feeding?

_____

_____

_____

61. A 42-year-old woman went on a self-imposed "starvation" diet. In addition to losing 120 pounds, her doctor says that her thymus has atrophied beyond what is expected for her age. What effect may this have on her immune system?

_____

_____

_____

pathophysiology

Follow the instructions for the statement.

62. Fill in the missing cells.

Disease	Etiology	Signs and Symptoms	Treatment
Systemic lupus erythematosis			
Chronic fatigue syndrome			

The Respiratory System

13

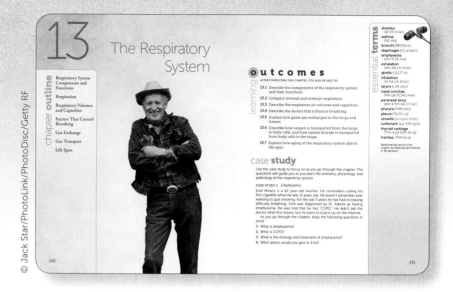

vocabulary review

MATCHING

Match the key terms in the right column with the definitions in the left column by writing the letter of the correct answer in the space provided.

_____ 1. A dual organ of the digestive and respiratory systems

_____ 2. The largest single cartilage of the larynx

_____ 3. The true vocal folds and the space between them

_____ 4. The airway extending from the larynx to the bronchi

_____ 5. The serous membrane that covers the lungs and lines the walls of the chest and diaphragm

_____ 6. The functional unit of the lungs

_____ 7. The major muscle of respiration

_____ 8. Lobed structures on the lateral walls of the nasal cavity that function to increase air turbulence

_____ 9. Mucus-lined air cavities in the skull

_____ 10. The "voice box"

_____ 11. The "tubes" that branch off the trachea and divide into smaller and smaller diameter airways

_____ 12. A COPD caused by cigarette smoking where there is destruction of alveoli and flattening of the diaphragm

_____ 13. Chemical secreted by special cells in the lungs to decrease surface tension and allow expansion of the lungs

_____ 14. Buildup of fluid in the pleural cavity

_____ 15. The number two cause of death in the United States

a. Alveolus
b. Bronchi
c. Diaphragm
d. Emphysema
e. Glottis
f. Larynx
g. Lung cancer
h. Nasal conchae
i. Paranasal sinuses
j. Pharynx
k. Pleura
l. Pleural effusion
m. Surfactant
n. Thyroid cartilage
o. Trachea

111

content review

MULTIPLE CHOICE

In the space provided, write the letter of the choice that best completes each statement or answers each question.

_____ 16. The upper respiratory tract consists of all the following *except*
   a. Pharynx
   b. Nose
   c. Paranasal sinuses
   d. Larynx

_____ 17. The lower respiratory tract consists of all the following *except*
   a. Pharynx
   b. Alveoli
   c. Bronchioles
   d. Trachea

_____ 18. The nose is made up of bone, adipose tissue, and _____ all of which is covered by skin.
   a. Elastic cartilage
   b. Hyaline cartilage
   c. Fibrocartilage
   d. Myelin cartilage

_____ 19. The _____ or nostrils are the openings of the nose.
   a. External nares
   b. Internal nares
   c. Nasal conchae
   d. Paranasal sinuses

_____ 20. The ethmoid and vomer bones make up the nasal
   a. Sinuses
   b. Conchae
   c. Septum
   d. Paranasal sinuses

_____ 21. What is the term for a paranasal sinus that consists of four to six small cavities?
   a. Sphenoid
   b. Maxillary
   c. Ethmoid
   d. Frontal

_____ 22. Which of the following is the largest of the paranasal sinuses?
   a. Frontal
   b. Maxillary
   c. Ethmoid
   d. Sphenoid

_____ 23. All of the following are regions of the pharynx *except*
   a. Linguopharynx
   b. Oropharynx
   c. Nasopharynx
   d. Laryngopharynx

_____ 24. What family of viruses is the cause of many upper respiratory infections?
   a. Cytomegalic virus
   b. Rhinovirus
   c. Leo virus
   d. Canine virus

_____ 25. How many regions make up the pharynx?
   a. Four
   b. Two
   c. Five
   d. Three

26. The trachea has about 20 C-shaped rings that
    a. Open anterior
    b. Open lateral
    c. Open posterior
    d. Are closed

27. All of the following are true of influenza *except*
    a. It is usually caused by virulent bacteria
    b. It is typically self-limiting
    c. It has killed millions of people over the course of history
    d. People with compromised immune systems are at greater risk of developing the flu

28. What is the term to describe the correct way to cover a cough?
    a. Flu blanket
    b. Flu muffle
    c. Flu salute
    d. Flu silencer

29. The point of bifurcation of the trachea into right and left main stem bronchi is called what?
    a. Hilum
    b. Carina
    c. Marina
    d. Nilum

30. In asthma, there is hyperactivity of the _____ in the airways.
    a. Cardiac muscle
    b. Skeletal muscle
    c. Smooth muscle
    d. Fat cells

31. Most cases of chronic bronchitis are due to
    a. Viral infections
    b. Chewing tobacco
    c. Cigarette smoking
    d. Bacterial infections

32. What is the name of the procedure that is performed to remove fluid and/or pus from the thorax?
    a. Choriocentesis
    b. Thracoectomy
    c. Amniocentesis
    d. Thoracocentesis

33. Approximately what percentage of primary cases of tuberculosis are asymptomatic?
    a. 90%
    b. 30%
    c. 50%
    d. 70%

34. _____ or collapsed lung can occur due to various reasons such as trauma or after thoracic surgery.
    a. Thoracostasis
    b. Atelectasis
    c. Mediastinitis
    d. Pleural effusion

35. Which of the following is an acute type of bacterial pneumonia caused by a Gram-negative organism that grows in standing water?
    a. Mycoplasma pneumonia
    b. Tuberculosis
    c. Legionnaire's disease
    d. Respiratory syncytial virus

FILL IN THE BLANKS

In the space provided, write the word or phrase that best completes each sentence. Not all words or phrases are used.

36. _____ is the coughing up of blood.

37. Through the process of _____, oxygen and carbon dioxide are exchanged across the thin capillary walls.

38. _____ is the sudden, unexpected, and unexplainable death of an infant under one year of age.

39. Tuberculosis is caused by the _____ organism.

40. _____ is the amount of air that remains in the lungs even after a forceful expiration.

41. The _____ controls the rhythm and depth of breathing.

42. The _____ controls the rate of breathing.

43. In the lungs the concentration of oxygen is _____ than the concentration of oxygen in the blood vessels.

44. At the cellular level, the concentration of carbon dioxide is _____ in the blood than in the tissues.

45. By the time someone reaches 80 years of age, he or she has about _____ the ventilation of a 20-year-old.

a. Diffusion
b. Greater
c. Half
d. Hematuria
e. Hemoptysis
f. Less
g. Medulla oblongata
h. *Mycobacterium leprae*
i. *Mycobacterium tuberculosis*
j. Osmosis
k. Pons
l. Residual volume
m. Respiratory distress syndrome
n. Sudden infant death syndrome

SHORT ANSWER

Write the answer to each statement on the lines provided.

46. Identify and describe the organs of the respiratory system.

_____

_____

_____

_____

114 UNIT 2 Concepts of Common Illness by System

47. Explain how gases are exchanged in the lungs.

48. Compare the structure and function of the two types of alveolar cells.

49. Describe the structure and function of the nasal conchae.

50. Describe and compare the right and left lungs.

LABELING

Follow the directions and write the answers on the lines provided.

51. In the following illustration showing the structures of the respiratory system, identify these terms by writing them on the lines provided: *bronchiole, bronchus, diaphragm, epiglottis, glottis, larynx, lung, nasal cavity, nostril, pharynx, trachea.*

a. _____

b. _____

c. _____

d. _____

e. _____

f. _____

g. _____

h. _____

i. _____

j. _____

k. _____

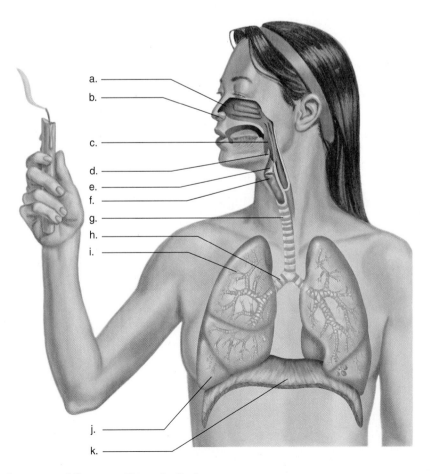

116   UNIT 2   Concepts of Common Illness by System

52. Using the following figures showing anterior and posterior views of the larynx, identify these terms by writing them on the lines provided: *cricoid cartilage, epiglottic cartilage, hyoid bone, thyroid cartilage, trachea.* (*Hint:* Each term is used twice.)

a. _____
b. _____
c. _____
d. _____
e. _____
f. _____
g. _____
h. _____
i. _____
j. _____

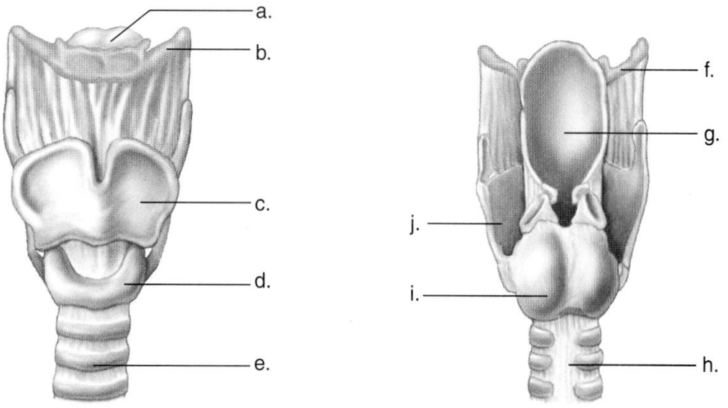

From *Hole's Human Anatomy & Physiology*, 12e, by Shier/Butler/and Lewis. Copyright © 2009. Reprinted by permission of McGraw-Hill Companies Inc.

critical thinking / application

CRITICAL THINKING

Write the answer to each question on the lines provided.

53. Why does it make sense that carbon dioxide, not oxygen, is the stimulus for breathing under most situations?

_____
_____
_____

CHAPTER 13   The Respiratory System

54. Why do you think drug-resistant strains of *Mycobacterium tuberculosis* have developed?

APPLICATION

Write the answer to the statement on the lines provided.

55. Someone you know has just been diagnosed with stage 3 lung cancer. Explain what this means in terms of treatment and prognosis.

case studies

Write your response to each case study question or statement on the lines provided.

56. A 32-year-old medial assistant is going to India on a medical mission trip. What should she know about tuberculosis?

57. A 14-year-old girl was brought in to the doctor's office because of sneezing, watery eyes, and runny nose. Describe what an allergy is and how these symptoms are consistent with an allergic response.

pathophysiology

Follow the instructions for the statement.

58. Fill in the missing cells.

Disease	Etiology	Signs and Symptoms	Treatment
Sinusitis			
Influenza			
Pleural effusion			
Pneumonoconiosis			

The Nervous System 14

vocabulary review

MATCHING

Match the key terms in the right column with the definitions in the left column by writing the letter of the correct answer in the space provided.

1. In charge of the body's automatic functions
2. Responsible for "rest and digest"
3. Its coating creates white matter
4. Contains ribosomes and mitochondria
5. The brain and spinal cord
6. Transmits electrical signals to the cell body
7. Releases neurotransmitters to provide transmission between neurons
8. Produces the "fight-or-flight" response
9. Consists of the nerves in the brainstem and sacral regions of the spinal cord
10. Carries action potentials away from the cell body

a. ANS
b. Axon
c. Cell body
d. CNS
e. Dendrite
f. Myelin sheath
g. Parasympathetic nervous system
h. PNS
i. Sympathetic nervous system
j. Synaptic knob

content review

MULTIPLE CHOICE

In the space provided, write the letter of the choice that best completes each statement or answers each question.

_____ 11. Which of the following neuroglial cells is a phagocyte?
   a. Oligodendrocyte
   b. Microglia
   c. Astrocyte
   d. Macroglia

_____ 12. Which lobe of the cerebrum contains the primary visual areas that interpret what a person sees?
   a. Temporal
   b. Occipital
   c. Frontal
   d. Parietal

_____ 13. Which of the following areas of the brain contains CSF?
   a. Gyri
   b. Blood-brain barrier
   c. Medulla oblongata
   d. Ventricles

_____ 14. Which of the following is a function of the cerebellum?
   a. The interpretation of hearing
   b. The regulation of breathing
   c. The coordination of complex skeletal muscle contractions needed for body movement
   d. The secretion of hormones

_____ 15. Which of the following acts as an "interpreter" for the other two types of neurons?
   a. Interneurons
   b. Microglia
   c. Neurotransmitters
   d. Myelin

_____ 16. Which procedure involves the use of a needle to remove CSF from the subarachnoid space?
   a. Cerebrocentesis
   b. MRI
   c. CT scan
   d. Lumbar puncture

_____ 17. Which cranial nerve does *not* control the eyes or visual functioning?
   a. II
   b. III
   c. IV
   d. V

_____ 18. Which of the following cells is found in the PNS?
   a. Astrocyte
   b. Schwann cell
   c. Oligodendrocyte
   d. Ependymal cell

_____ 19. The decision-making function of the nervous system takes place in the
   a. Cerebrum
   b. Cerebellum
   c. Pons
   d. Brain stem

_____ 20. What is another name for sensory neurons?
   a. Efferent neurons
   b. Interneurons
   c. Afferent neurons
   d. Association neurons

21. What is the term for the middle layer of the meninges?
    a. Arachnoid mater
    b. Pia mater
    c. Dura mater
    d. Intermediate mater

22. What is the largest part of the brain?
    a. Cerebrum
    b. Cerebellum
    c. Pons
    d. Medulla oblongata

23. The cranial nerve that innervates the lateral rectus muscle of the eye is called the
    a. Abducens
    b. Oculomotor
    c. Optic
    d. Trochlear

24. A group of nerve cell bodies outside the CNS is called a
    a. Ganglion
    b. Nerve tract
    c. Nuclei
    d. Nerve

25. Which cranial nerve is responsible for transmitting aromas to the brain for interpretation?
    a. Optic
    b. Olfactory
    c. Trigeminal
    d. Facial

26. Which of the following is the area of skin innervated by a single spinal or cranial nerve?
    a. Dermatome
    b. Sclerotome
    c. Myotome
    d. Rhizome

27. The outermost layer of the cerebrum is called the
    a. Myelin
    b. Basal ganglia
    c. Cortex
    d. Cerebellum

28. What type of chemicals is released from the synaptic knob?
    a. Neurotransmitters
    b. Enzymes
    c. Vitamins
    d. Phospholipids

29. Which lobe of the brain contains auditory areas that interpret sound?
    a. Occipital
    b. Frontal
    c. Temporal
    d. Parietal

30. Which part of the brain is involved in the regulation of blood pressure?
    a. Thalamus
    b. Epithalamus
    c. Pons
    d. Hypothalamus

FILL IN THE BLANKS

In the space provided, write the word or phrase that best completes each sentence. Not all words or phrases are used.

31. _____ are the neuroglial cells that anchor blood vessels to nerves.

32. The _____ is formed by a tight junctions of capillaries that protect the delicate tissue of the CNS.

33. The myelin sheath is created by the _____.

34. _____ are found around some axons in the CNS, producing what is known as the myelin sheath.

35. Neuron cell membranes are polarized; the term for this is _____.

36. Unmyelinated axons of the CNS are referred to as _____.

37. The _____ tracts of the spinal cord carry sensory information to the brain.

38. A predictable automatic response to a stimulus is called a(n) _____.

39. The cerebrum is split into four sections called _____.

40. The _____ consists of the midbrain, the pons, and the medulla oblongata.

41. _____ neuron is another name for sensory neuron.

42. A(n) _____ uses contrast material so that the blood vessels in the brain can be visualized.

43. The _____ enlargement of the spinal cord contains motor neurons that control the muscles of the legs.

44. _____ neurons carry information from the periphery to the CNS.

45. The _____ is composed of the sympathetic and parasympathetic nervous systems.

a. Afferent
b. Ascending
c. Astrocytes
d. Autonomic nervous system
e. Blood-brain barrier
f. Brainstem
g. Cerebellum
h. Cerebral angiography
i. Descending
j. Gray matter
k. Lobes
l. Lobules
m. Lumbar
n. Microglia
o. Neural reticulum
p. Oligodendrocytes
q. Reflex
r. Resting
s. Sensory
t. Somatic nervous system
u. White matter

SHORT ANSWER

Follow the directions and write the answer to each statement on the lines provided.

46. Describe the components of a neuron.

_____

_____

_____

_____

47. Describe the difference between a gyrus and a sulcus.

_____

_____

_____

_____

48. Fill in the missing cells with information about the cranial nerves.

Cranial Nerve Number	Name of Cranial Nerve	Function of Cranial Nerve
II		
IV		
V		
VII		
X		

49. Give the correct number for each of the following:

   a. Total number of spinal nerves _____

   b. Total number of cervical spinal nerves _____

   c. Total number of thoracic spinal nerves _____

   d. Total number of lumbar spinal nerves _____

   e. Total number of sacral spinal nerves _____

   f. Total number of coccygeal spinal nerves _____

LABELING

Follow the directions and write the answers on the lines provided.

50. Using the following illustration of a synapse, identify these terms by writing them on the lines provided: *axon membrane, dendrite of postsynaptic neuron, depolarized membrane, mitochondrion, neurotransmitter, polarized membrane, presynaptic neuron, synaptic cleft, synaptc knob. synaptic vesicle, vesicle releasing neurotransmitter.*

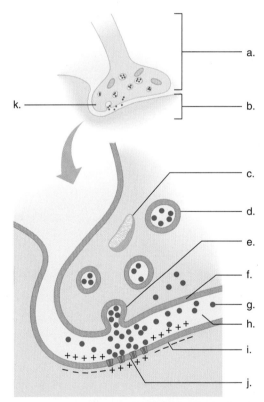

From *Hole's Human Anatomy & Physiology,* 12e, by Shier/Butler/and Lewis. Copyright © 2009. Reprinted by permission of McGraw-Hill Companies Inc.

a. _____

b. _____

c. _____

d. _____

e. _____

f. _____

g. _____

h. _____

i. _____

j. _____

k. _____

51. Using the following illustration of a reflex arc, identify these terms by writing them on the lines provided: *axon of motor neuron, axon of sensory neuron, cell body of motor neron, cell body of sensory neuron, effector-quadriceps muscle, patella, patellar ligament, receptor ends of sensory neuron, spinal cord.*

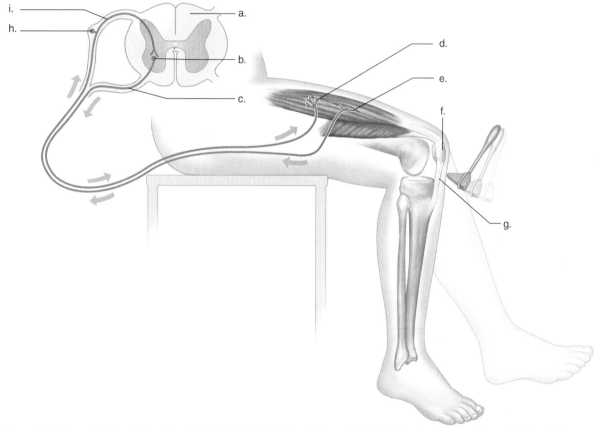

From *Hole's Human Anatomy & Physiology*, 12e, by Shier/Butler/and Lewis. Copyright © 2009. Reprinted by permission of McGraw-Hill Companies Inc.

CHAPTER 14 The Nervous System

a. _____
b. _____
c. _____
d. _____
e. _____
f. _____
g. _____
h. _____
i. _____

critical thinking / application

CRITICAL THINKING

Write the answer to each statement on the lines provided.

52. Explain the differences between tension and migraine headaches.

_____
_____
_____
_____
_____

53. Explain the major differences between the somatic nervous system and autonomic nervous system.

_____
_____
_____
_____
_____

APPLICATION

Write the answers to each statement on the lines provided.

54. Nerve Plexuses. Describe the nerves branching off each of the following plexuses:

   a. Cervical _____

   b. Brachial _____

   c. Lumbosacral _____

55. Reflex Testing. Name the nerves that may be damaged if the following reflexes are abnormal:

   a. Biceps reflex _____

   b. Knee reflex _____

   c. Abdominal reflexes _____

case studies

Write your response to each case study question on the lines provided.

56. A patient has bacterial meningitis. What are the meninges and what is their function?

_____

_____

_____

_____

57. Your patient shows increased electrical activity within the brain on an EEG. What condition may cause this finding?

_____

_____

_____

_____

pathophysiology

Follow the instructions for the statement.

58. Fill in the missing cells.

Disease	Etiology	Signs and Symptoms	Treatment
Parkinson's disease			
Cerebrovascular accident			
Migraines			

The Urinary System 15

vocabulary review

MATCHING

Match the key terms in the right column with the definitions in the left column by writing the letter of the correct answer in the space provided.

_____ 1. The triangle formed by the three openings within the bladder

_____ 2. An alternate term for urination

_____ 3. Small arteries that bring blood to the glomerulus

_____ 4. The hormone that assists with red blood cell formation

_____ 5. Tightly coiled capillaries of the nephron that begin the filtration process

_____ 6. The tube that carries urine from the bladder to the outside of the body

_____ 7. The enzyme that converts angiotensinogen to angiotensin I

_____ 8. The functional unit of the kidney

_____ 9. The position of the kidneys

_____ 10. The tubes that carry urine from the kidney to the urinary bladder

a. Afferent arterioles
b. Efferent arterioles
c. Erythropoietin
d. Glomerulus
e. Loop of Henle
f. Micturition
g. Nephron
h. Pyelonephritis
i. Renal pelvis
j. Renin
k. Retroperitoneal
l. Trigone
m. Ureter
n. Urethra

131

content review

MULTIPLE CHOICE

In the space provided, write the letter of the choice that best completes each statement or answers each question.

_____ 11. Which of the following is the cuplike structure inside the kidney?
   a. Pelvis
   b. Calyce
   c. Column
   d. Ureter

_____ 12. The triangular-shaped areas of the renal medulla are the
   a. Renal pelvis
   b. Renal columns
   c. Trigones
   d. Renal pyramids

_____ 13. Which of the following is *not* a renal tubule?
   a. The glomerular capsule
   b. The distal convoluted tubule
   c. The loop of Henle
   d. The proximal convoluted tubule

_____ 14. Which of the following delivers blood to the glomerulus?
   a. The afferent arterioles
   b. The efferent arterioles
   c. The interlobular arteries
   d. The renal vein

_____ 15. Which of the following is responsible for tubular reabsorption?
   a. The ureters
   b. The collecting tubules
   c. The Bowman's capsule
   d. The proximal convoluted tubules

_____ 16. The three openings of the floor of the bladder are
   a. One opening for the ureter and two for each urethra
   b. One opening for the urethra and two for the ureters
   c. One opening for the urethra and two for the urinary arteries
   d. One opening for the ureter and two for the urinary arteries

_____ 17. The trigone refers to
   a. The proximal convoluted tubule, the loop of Henle, and the distal convoluted tubule
   b. The two ureters and the urethra
   c. The "jobs" of the kidney, which are filtration, reabsorption, and secretion
   d. The area of the bladder that encompasses the ureteral and urethral openings

_____ 18. Secretion refers to
   a. The formation and release of erythropoietin
   b. The formation of filtrate
   c. The removal of urine from the bladder
   d. A substance moving out of the peritubular capillaries and into the renal tubules

_____ 19. What is the inability of kidneys to create urine called?
   a. Renal failure
   b. Incontinence
   c. Retention
   d. Renal calculi

_____ 20. The detrusor muscle is responsible for
   a. Moving urine from the ureters to the bladder
   b. Retaining urine in the bladder
   c. Expelling urine from the bladder
   d. Sensing that the bladder is full

21. The kidneys' position in the body is described as
    a. Retroperitoneal
    b. Lateral to the heart
    c. Medial to the heart
    d. Mediastinal

22. Which of the following is the deepest layer of tissue surrounding each kidney?
    a. Bowman's capsule
    b. Renal fascia
    c. The renal capsule
    d. Renal fat

23. Which of the following helps the kidneys regulate blood volume and blood pressure?
    a. Epinephrine
    b. Oxytocin
    c. Growth hormone
    d. Renin

24. What is the approximate mass of the kidneys in relation to the body's total mass?
    a. 10%
    b. 1%
    c. 15%
    d. 20%

25. The renal veins drain into the
    a. Inferior vena cava
    b. Superior vena cava
    c. Aorta
    d. Right subclavian vein

26. The middle section of the renal tubule is called
    a. Bowman's capsule
    b. The proximal convoluted tubule
    c. The distal convoluted tubule
    d. The loop of Henle

27. A sudden loss of kidney function is known as
    a. Diabetes mellitus
    b. Chronic renal failure
    c. Acute renal failure
    d. Uremia

28. What is the term for blood in the urine?
    a. Anuria
    b. Oliguria
    c. Pyuria
    d. Hematuria

29. The rate of filtration is largely controlled by
    a. The sympathetic nervous system
    b. The parasympathetic nervous system
    c. The somatic nervous system
    d. The basal ganglia

30. Urine normally contains all of the following *except*
    a. Uric acid
    b. Urea
    c. Glucose
    d. Ions

FILL IN THE BLANKS

In the space provided, write the word or phrase that best completes each sentence. Not all words or phrases are used.

31. The innermost layer of the kidney is called the renal _____.

32. A nephron consists of a renal corpuscle and a(n) _____.

33. Rhythmic muscular contractions known as _____ propel urine through the ureters.

34. Erythropoietin causes the formation of _____.

35. _____ is a condition in which the kidneys slowly lose their ability to function.

36. _____ is an inflammation of the glomeruli.

37. _____ is a disorder in which the kidneys enlarge because of the presence of many cysts within them.

38. The smooth muscle layer of the urinary bladder is known as the _____.

39. _____ is more commonly known as a bladder infection.

40. _____ describes the condition in which "stones" are found within the kidney.

a. Acute renal failure
b. Cholelithiasis
c. Chronic renal failure
d. Cystitis
e. Detrusor muscle
f. Glomerulonephritis
g. Medulla
h. Peristalsis
i. Polycystic kidney disease
j. Pyelonephritis
k. Red blood cells
l. Renal calculi
m. Tubular system
n. Urethritis

SHORT ANSWER

Write the answer to each statement on the lines provided.

41. Describe the major events of micturition.

_____
_____
_____
_____
_____

42. Describe the characteristics of the layers of the urinary bladder.

43. Explain the steps women can take to prevent cystitis.

44. Describe how Kegel exercises are performed.

45. Describe the blood and nerve supply to the kidneys.

LABELING

Follow the directions and write the answers on the lines provided.

46. Using the following illustration showing the organs of the urinary system, identify these terms by writing them on the lines provided: *aorta, bladder, hilum, inferior vena cava, kidney, renal artery, renal vein, ureters, urethra.*

a. _____

b. _____

c. _____

d. _____

e. _____

f. _____

g. _____

h. _____

i. _____

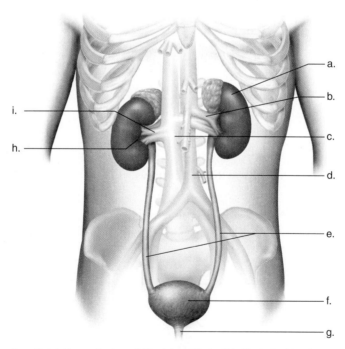

From *Hole's Human Anatomy & Physiology,* 12e, by Shier/Butler/and Lewis. Copyright © 2009. Reprinted by permission of McGraw-Hill Companies Inc.

47. Using the following figure showing a section of a kidney, identify these terms by writing them on the lines provided: *major calyx, minor calyx, renal capsule, renal column, renal cortex, renal medulla, renal papilla (tip of renal pyramid), renal pelvis, renal pyramid, renal sinus, ureter.*

a. _____

b. _____

c. _____

d. _____

e. _____

f. _____

g. _____

h. _____

i. _____

j. _____

k. _____

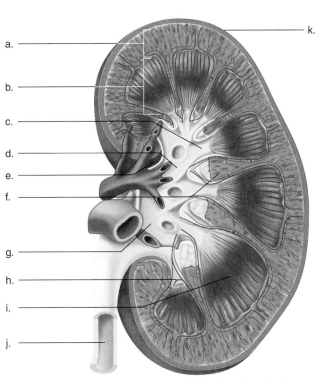

From *Hole's Human Anatomy & Physiology*, 12e, by Shier/Butler/and Lewis. Copyright © 2009. Reprinted by permission of McGraw-Hill Companies Inc.

critical **thinking** / **application**

CRITICAL THINKING

Write the answer to each statement on the lines provided.

48. Explain how hemorrhage can lead to renal failure.

49. Explain why the right renal artery is longer than the left renal artery.

APPLICATION

Follow the directions for the application.

50. Explain how low blood pressure affects glomerular filtration.

case studies

Write your response to each case study question or statement on the lines provided.

51. A 28-year-old female recently gave birth to twin girls. She says she now has a problem with leakage of urine. What is the cause of this problem and what are some of the treatments that may be beneficial?

_____

_____

_____

_____

52. A young woman has a history of recurrent urinary tract infections (UTI). Explain why females have more UTIs than men and what can she do to decrease the incidence of them.

_____

_____

_____

_____

pathophysiology

Follow the instructions for the statement.

53. Fill in the missing cells.

Disease	Etiology	Signs and Symptoms	Treatment
Polycystic kidney disease			
Glomerulonephritis			
Pyelonephritis			

CHAPTER 15   The Urinary System

The Male Reproductive System 16

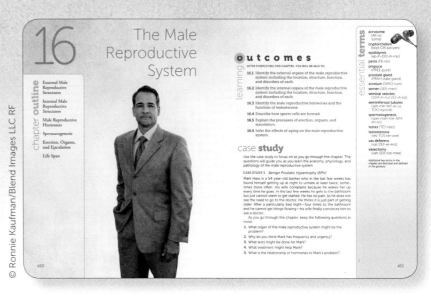

vocabulary review

MATCHING

Match the key terms in the right column with the definitions in the left column by writing the letter of the correct answer in the space provided.

_____ 1. The pouch of skin that holds the testes
_____ 2. The male gonads
_____ 3. A male sex hormone needed for the production of sperm
_____ 4. A condition in which a testicle did not descend during infancy
_____ 5. Tightly coiled tubes located in the testes where sperm are produced
_____ 6. Another term for the foreskin that covers the glans penis
_____ 7. Organ that surrounds the proximal end of the penis and secretes a milky alkaline fluid
_____ 8. A coiled tube on top of each testis where spermatids mature into sperm
_____ 9. A tubular structure connected to each epididymis that carries sperm cells to the urethra
_____ 10. Saclike organs that secrete an alkaline fluid rich in sugars and prostaglandins
_____ 11. A mixture of sperm cells and fluids from various glands
_____ 12. The process of sperm cell formation from spermatogonia
_____ 13. An enzyme-filled sac that covers the head of a sperm cell
_____ 14. Also known as the bulbourethral glands
_____ 15. The cutting and tying of the vas deferens to prevent sperm from reaching the ovum

a. Acrosome
b. Cryptorchidism
c. Cowper's glands
d. Epididymis
e. Prepuce
f. Prostate gland
g. Scrotum
h. Semen
i. Seminal vesicles
j. Seminiferous tubules
k. Spermatogenesis
l. Testes
m. Testosterone
n. Vas deferens
o. Vasectomy

content review

MULTIPLE CHOICE

In the space provide, write the letter of the choice that best completes each statement or answers each question.

_____ 16. The portion of a sperm cell that contains the genetic information from the biological father is the
   a. Tail
   b. Head
   c. Midpiece
   d. Acrosome

_____ 17. For viability of the sperm, what should the temperature of the scrotum be?
   a. One degree lower than the core of the body
   b. One degree higher than the core of the body
   c. Ten degrees lower than the core of the body
   d. Ten degrees higher than the core of the body

_____ 18. Testosterone is produced by
   a. The pituitary gland
   b. The testes
   c. The hypothalamus
   d. The pineal gland

_____ 19. Which of the following is *not* true of testicular cancer?
   a. It more commonly occurs in older men
   b. It is more aggressive than prostate cancer
   c. Cryptorchidism is a predisposing factor
   d. A hard, painless lump is a common early symptom

_____ 20. A symptom of epididymitis is dysuria which means
   a. Pus in the urine
   b. Blood in the urine
   c. Painful urination
   d. Inability to urinate

_____ 21. The seminal vesicles secrete an alkaline fluid rich in sugars and prostaglandins that make up approximately what percentage of the semen volume?
   a. 20%
   b. 40%
   c. 60%
   d. 80%

_____ 22. What is the most common cancer in men older than age 40?
   a. Breast cancer
   b. Colon cancer
   c. Testicular cancer
   d. Prostate cancer

_____ 23. All of the following are true of prostate cancer *except*
   a. It is more common in younger men
   b. It is not as aggressive as testicular cancer
   c. Testosterone can spur its growth
   d. The exact cause is unknown

_____ 24. All of the following are true of prostatitis *except*
   a. It can be caused by bacterial infections
   b. It can be caused by frequent urination
   c. Acute or chronic dysuria may be a symptom
   d. Weight loss is a symptom

_____ 25. Which of the following is true of benign prostatic hypertrophy?
   a. Its incidence decreases as men get older
   b. Dihydrotestosterone (DHT) inhibits the growth of the prostate gland

c. Diagnosis is most commonly done by x-ray

d. Transurethral resection of the prostate (TURP) is not successful in the treatment of BPH

____ 26. A normal sperm count is
   a. Less than 40 million sperm per milliliter of ejaculate
   b. Between 40 and 60 million sperm per milliliter of ejaculate
   c. Between 60 and 80 million sperm per milliliter of ejaculate
   d. More than 80 million sperm per milliliter of ejaculate

____ 27. Which of the following is *not* a risk factor for prostate cancer?
   a. Eating a diet high in lycopene
   b. Being African American
   c. Having first-degree relatives with prostate cancer
   d. Advancing age

____ 28. How many chromosomes do spermatogonia contain?
   a. 23
   b. 44
   c. 46
   d. 92

____ 29. The end result of meiosis in the male is
   a. Two spermatids
   b. Four spermatids
   c. Six spermatids
   d. Eight spermatids

____ 30. A mature sperm has all of the following *except*
   a. Cilia
   b. A tail
   c. A midpiece
   d. A head

____ 31. The combination of the sperm and secretions into the urethra is called
   a. Emission
   b. Ejaculation
   c. Transmission
   d. Meiosis

____ 32. All of the following are true of impotence or erectile dysfunction *except*
   a. Most cases are due to psychological factors
   b. It is the inability to maintain an erection and complete intercourse
   c. It increases with age
   d. Diabetes may be a cause of the disorder

____ 33. A 62-year-old man has been diagnosed with BPH. What does this mean?
   a. He has an undescended testicle
   b. He has prostate cancer
   c. He has nonmalignant enlargement of the prostate gland
   d. He has erectile dysfunction

____ 34. Which statement is accurate regarding the hormones related to BPH?
   a. Decreased DHT
   b. Decreased estrogen
   c. Increased testosterone
   d. Increased DHT

____ 35. All of the following are true of the testes *except*
   a. It is abnormal to have one testicle lower in position than the other
   b. They produce testosterone
   c. They produce sperm cells
   d. They begin their development in the abdominopelvic cavity of the fetus

____ 36. Testosterone is produced by what type of cells?
   a. Sperm cells
   b. Interstitial cells
   c. Ependymal cells
   d. Fibroblasts

FILL IN THE BLANKS

In the space provided, write the word or phrase that best completes each sentence. Not all words or phrases are used.

37. The _____ is the muscular gland located at the base of the male urethra.

38. The _____ holds the testes away from the rest of the body, keeping their temperature one degree lower than the core of the body.

39. _____ cells give rise to sperm cells.

40. The _____ or foreskin covers the glans penis in an uncircumcised penis.

41. Screening tests such as DRE and _____ are used to detect prostate cancer.

42. Prostatitis may cause a low sperm count which is known as _____.

43. _____ refers to nonmalignant enlargement of the prostate gland.

44. _____ is a procedure that may be performed to remove the enlarged tissue in BPH.

45. The _____ releases gonadotropin releasing hormone (GnRH).

46. Follicle-stimulating hormone is secreted by the _____.

a. Anterior pituitary gland
b. Benign prostatic hypertrophy
c. Hemoptysis
d. Hypothalamus
e. Oligospermia
f. Posterior pituitary gland
g. Prepuce
h. Prostate
i. Prostate cancer
j. Prostate specific antigen (PSA)
k. Scrotum
l. Sertoli cells
m. Spermatogenic cells
n. Testes
o. Transurethral resection of the prostate (TURP)

SHORT ANSWER

Write the answer to each statement on the lines provided.

47. Describe how sperm cells are formed.

_____
_____
_____
_____

48. Describe the actions of testosterone.

49. Describe how a testicular self-examination is to be performed.

50. Describe the structure of a mature sperm.

51. Explain the process of erection and ejaculation.

LABELING

Follow the directions and write the answers on the lines provided.

52. Using the following illustration of the male reproductive system, identify these terms by writing them on the lines provided: *ductus (vas) deferens, penis, prostate gland, scrotum, seminal vesicle, testis.*

a. _____

b. _____

c. _____

d. _____

e. _____

f. _____

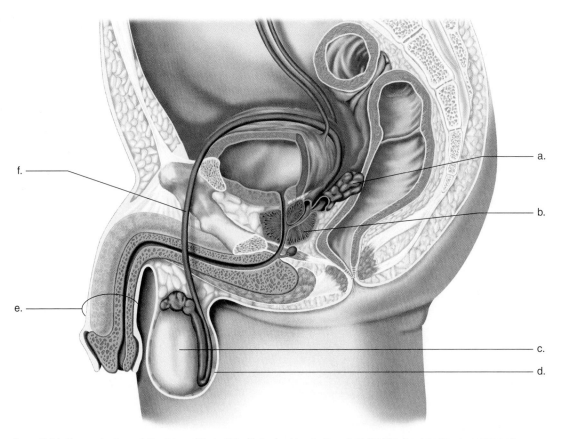

From *Hole's Human Anatomy & Physiology*, 12e, by Shier/Butler/and Lewis. Copyright © 2009. Reprinted by permission of McGraw-Hill Companies Inc.

146 UNIT 2 Concepts of Common Illness by System

critical thinking / application

CRITICAL THINKING

Write the answer to each question on the lines provided.

53. Smoking may have an effect on the functioning of cilia and flagella. Why do you think this might cause sterility in men?

54. Based on your knowledge of the male reproductive organs, why do you think BPH causes difficulty with urination?

APPLICATION

Write the answer to the statement on the lines provided.

55. Explain why regular prostate screening is important for men to have done.

case studies

Write your response to each case study question on the lines provided.

56. A 58-year-old male chemist has noticed blood in his urine lately. He has had no trauma and he has been a heavy cigarette smoker for 41 years. Does he have any risk factors for prostate cancer? What is the treatment for this type of disease?

_____

_____

_____

_____

pathophysiology

Follow the instructions for the statement.

57. Fill in the missing cells.

Disease	Etiology	Signs and Symptoms	Treatment
Epididymitis			
Prostate cancer			
Erectile dysfunction			

The Female Reproductive System   17

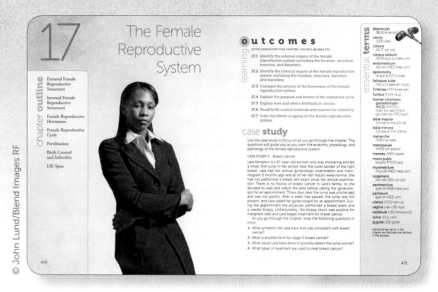

vocabulary review

MATCHING

Match the key terms in the right column with the definitions in the left column by writing the letter of the correct answer in the space provided.

_____ 1. Formation of the female gametes
_____ 2. Oviduct, the tube that runs from the uterus to the ovary
_____ 3. The hollow, muscular organ in the female that is the site of menstruation and implantation of the embryo and development of the fetus
_____ 4. The mucous membrane lining of the uterus
_____ 5. The external genitalia of the female
_____ 6. A woman's menstrual phase
_____ 7. The inferior constricted part of the uterus
_____ 8. The smooth muscle layer of the uterus
_____ 9. The termination of the menstrual cycles
_____ 10. The fertilized ovum
_____ 11. The first menses
_____ 12. Rounded fatty prominence over the pubic symphysis
_____ 13. A hollow ball of cells in the development of the embryo
_____ 14. An erectile organ of the female, homologous to the male penis
_____ 15. The part of the uterus furthest from the os
_____ 16. The serosa of the uterus
_____ 17. The small space at the beginning of the vagina
_____ 18. The pelvic floor, the space between the anus and the vulva in the female and the anus and the scrotum in the male
_____ 19. The muscular, tubular organ that extends from the uterus to the vestibule
_____ 20. Fingerlike structures near the lateral ends of the fallopian tubes

a. Blastocyst
b. Cervix
c. Clitoris
d. Endometrium
e. Fallopian tubes
f. Fimbriae
g. Fundus
h. Menarche
i. Menopause
j. Menses
k. Mons pubis
l. Myometrium
m. Oogenesis
n. Perimetrium
o. Perineum
p. Uterus
q. Vagina
r. Vestibule
s. Vulva
t. Zygote

149

content review

MULTIPLE CHOICE

In the space provided, write the letter of the choice that best completes each statement or answers each question.

_____ 21. Which of the following are the folds of skin between the labia majora?
   a. The clitoris
   b. The mons pubis
   c. The labia minora
   d. The vagina

_____ 22. Which of the following are benign (noncancerous) tumors that grow in the uterine wall?
   a. Cervicitis
   b. Fibroids
   c. Endometriosis
   d. Dyspareunia

_____ 23. What is sometimes "clipped" during the birth process to avoid tearing?
   a. Mons pubis
   b. Clitoris
   c. Perineum
   d. *Treponema pallidum*

_____ 24. Which structure is anterior to the urethral meatus and contains the female erectile tissue and is rich in sensory nerves?
   a. Perineum
   b. Clitoris
   c. Episiotomy
   d. Vulva

_____ 25. The walls of the uterus and fallopian tubes contract to help propel sperm toward the upper ends toward the fallopian tubes
   a. During exercise
   b. During the birth process
   c. During menstruation
   d. During orgasm

_____ 26. The vulva includes which of the following structures?
   a. Clitoris, perineum, and mons pubis
   b. Clitoris and perineum pubis
   c. Perineum and mons pubis
   d. Clitoris and mons pubis

_____ 27. An inflammation of the vulva and vagina is called
   a. Fimbriae
   b. Vaginitis
   c. Vulvovaginitis
   d. Dimpling

_____ 28. Among the internal genitalia of the female are the
   a. Progesterone
   b. Ovaries
   c. Ovulation
   d. Vulva

_____ 29. What disease or diseases include AIDS, genital warts, gonorrhea, herpes simplex 1 and 2, pubic lice, and syphilis?
   a. ATP
   b. STI
   c. HIV
   d. Endometriosis

_____ 30. What are considered accessory organs of both the reproductive and integumentary systems?
   a. Fallopian tubes
   b. Bartholin's glands
   c. Female breasts
   d. Lungs

___ 31. More than 60% of the women in the United States between the ages of 30 and 50 have which of the following conditions?
   a. Fibrocystic breast change
   b. Breast cancer
   c. Ovarian cancer
   d. Oral cancer

___ 32. The hypothalamus secretes increasing amounts of GnRH beginning
   a. In puberty
   b. During the birth process
   c. During menopause
   d. During urination

___ 33. A collection of symptoms that occur just before the menstrual period is known as
   a. DES
   b. SAT
   c. PTH
   d. PMS

___ 34. The condition of having a developing offspring in the uterus is commonly called
   a. Ovulation
   b. Birth
   c. Pregnancy
   d. Cervicitis

___ 35. What is another term for contraception?
   a. Radiation
   b. Metaplasia
   c. Birth control
   d. Vaginitis

___ 36. The inability to conceive is also referred to as
   a. External genitalia
   b. Infertility
   c. Inflammation of the vagina
   d. Labia minora

___ 37. When a female experiences menarche, she has reached
   a. Her reproductive years
   b. Age 21
   c. Menopause
   d. The birth process

___ 38. The number of follicles and the production of estrogen decreases
   a. With exercise
   b. With the use of vitamins
   c. With an increase in fatty acids
   d. With age

___ 39. Estrogen and progesterone are responsible for the development of which of the following?
   a. Ova
   b. Mammary glands
   c. Ovarian cysts
   d. Endometriosis

___ 40. Which fatty area overlies the pubic symphysis?
   a. Perineum
   b. Labia majora
   c. Mons pubis
   d. Episiotomy

___ 41. What is the inflammation of the vagina called?
   a. Vaginitis
   b. Vestibule
   c. Bartholin's glands
   d. Oxytocin (OT)

___ 42. Which microscopic test looks at abnormal cervical cells that have been smeared onto a microscopic slide?
   a. Mastectomy
   b. Papanicolaou smear
   c. Lumpectomy
   d. FSH

CHAPTER 17 The Female Reproductive System

_____ 43. Which are female hormones?
   a. FSH and LH
   b. Estrogen and progesterone
   c. LH and corpus luteum
   d. FSH and hCG

_____ 44. What is the term for the inner region of the ovary?
   a. Cortex
   b. Lymphatic vessel
   c. Oocyte
   d. Medulla

_____ 45. Which condition causes severe menstrual cramps that limit daily activities?
   a. Dysmenorrhea
   b. Diarrhea
   c. Indigestion
   d. Dyspareunia

FILL IN THE BLANKS

In the space provided, write the word or phrase that best completes each sentence. Not all words or phrases are used.

46. Infection with human papillomavirus (HPV) that is transmitted sexually is the greatest single risk factor for _____.

47. About _____ of the infertility problems are in the female.

48. Generally, infertility in men is _____.

49. The outer region of the _____ is called the cortex.

50. The medulla contains _____.

51. Before a female is born, _____ develop in her ovarian cortex.

52. A female _____ make primary oocytes throughout their entire life.

53. _____ is considered more deadly than the other types of gynecological cancers.

54. The process of ovum formation is _____.

55. HPV that is _____ is the greatest single risk factor for cervical cancer.

56. Uterine cancer is most common in _____ women.

57. Primodial follicles contain a large cell called a primary oocyte and smaller cells called _____.

a. Adipose tissue
b. Age related
c. Cervical cancer
d. Does
e. Does not
f. Follicular cells
g. Human papillomavirus
h. Nerves, lymphatic vessels, and blood vessels
i. Not age related
j. One-third
k. Oogenesis
l. Ovarian cancer
m. Ovary
n. Postmenopausal
o. Premenopausal
p. Primary oocyte
q. Primordial follicles
r. Secondary oocyte
s. Two-thirds
t. Unite with the ovum

SHORT ANSWER

Write the answer to each statement on the lines provided.

58. Explain how and when ovulation occurs.

___

59. List the purpose and events of the menstrual cycle.

___

60. Explain how fertilization occurs.

___

61. List the causes of and treatments for infertility.

___

62. Identify the signs and symptoms of the silent killer.

___

LABELING

Follow the directions and write the answers on the lines provided.

63. Using the following illustration of the female reproductive and urinary systems, identify these terms by writing them on the lines provided: *cervix, clitoris, fallopian tubes, fimbriae, labia minus, labia majus, ovary, urethra, uterus, vagina.*

a. _____

b. _____

c. _____

d. _____

e. _____

f. _____

g. _____

h. _____

i. _____

j. _____

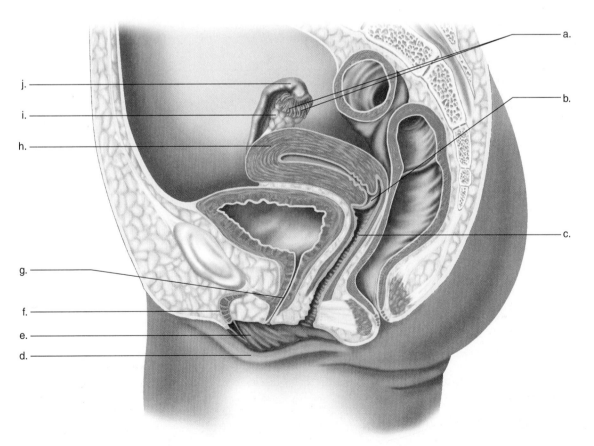

From *Hole's Human Anatomy & Physiology*, 12e, by Shier/Butler/and Lewis. Copyright © 2009. Reprinted by permission of McGraw-Hill Companies Inc.

critical thinking / application

CRITICAL THINKING

Write the answer to each statement on the lines provided.

64. Compare the actions of estrogen and progesterone.

65. Explain the purpose and events of the menstrual cycle.

APPLICATION

Write the answer to the statement on the lines provided.

66. List several birth control methods and explain why each is effective.

case studies

Write your response to each case study question on the lines provided.

67. A 44-year-old woman was recently diagnosed with ovarian cancer. Why is ovarian cancer harder to detect in its early stages than other types of reproductive cancers?

_____

_____

_____

_____

68. A 17-year-old sexually active teenager thinks because she takes a hot shower after having sex that this will protect her from STIs. What should she be told?

_____

_____

_____

_____

pathophysiology

Follow the instructions for the statement.

69. Fill in the missing cells.

Disease	Etiology	Signs and Symptoms	Treatment
Ovarian cancer			
Fibroids			
Vaginitis			
Breast cancer			

Human Development and Genetics

vocabulary review

MATCHING

Match the key terms in the right column with the definitions in the left column by writing the letter of the correct answer in the space provided.

____ 1. A solid sphere of cells produced by successive divisions of the fertilized egg

____ 2. The primary germ layer that gives rise to the nervous system, epidermis, and its derivatives

____ 3. An extraembryonic membrane that transfers nutrients to the embryo and is the site of blood cell formation for the embryo

____ 4. A fetal blood vessel that allows the blood to bypass the liver

____ 5. A structure that allows nutrients and oxygen from maternal blood to pass to embryonic blood

____ 6. The actual childbirth stage of the birth process

____ 7. The genetic makeup of an individual

____ 8. A prenatal test that may be used to diagnose Down syndrome and is performed early in the second trimester by withdrawing and testing some of the amniotic fluid surrounding a fetus

____ 9. A hormone that stimulates the enlargement of the mammary glands

____ 10. Comes from the corpus luteum, inhibits uterine contractions, and relaxes the ligaments of the pelvis in preparation for childbirth

____ 11. A segment of DNA that determines a body trait found in the human body

____ 12. The observable expression of the genotype

____ 13. The middle primary germ layer that gives rise to connective tissue, blood vessels, and muscle

____ 14. An infant the first four weeks after birth

____ 15. A connection between the pulmonary trunk and the aorta that allows blood to bypass the lungs

____ 16. The protective fluid-filled sac that surrounds the embryo

____ 17. The thinning and softening of the cervix during the birthing process

____ 18. An opening in the fetal heart that is located between the right and left atria, allowing blood to bypass the lungs

____ 19. The structure that develops during the embryonic stage that allows the exchange of substance between the fetus and mother and acts as part of the immune system

____ 20. A primary germ layer of the developing embryo that gives rise to the gastrointestinal tract and the respiratory tract

a. Amniocentesis
b. Amnion
c. Chorion
d. Ductus arteriosus
e. Ductus venosus
f. Ectoderm
g. Effacement
h. Endoderm
i. Foramen ovale
j. Gene
k. Genotype
l. Lactogen
m. Mesoderm
n. Morula
o. Neonate
p. Parturition
q. Phenotype
r. Placenta
s. Relaxin
t. Yolk sac

content review

MULTIPLE CHOICE

In the space provided, write the letter of the choice that best completes each statement or answers each question.

_____ 21. The newborn urinates a lot because the kidneys are too immature
   a. To produce any urine at all
   b. To work correctly
   c. To work without the mother's kidneys
   d. To concentrate the urine

_____ 22. The normal joining of the male sperm and the female ovum produces a zygote with
   a. 23 chromosomal pairs
   b. 18 chromosomes
   c. 46 chromosomal pairs
   d. No chromosomal pairs

_____ 23. What chromosome combination will a female child have that a male child does *not* have?
   a. Two X chromosomes
   b. One X and one Y chromosome
   c. Two Y chromosomes
   d. One A and one B chromosome

_____ 24. Which stage of the birth process normally lasts for 8 to 24 hours?
   a. Conception
   b. Fertilization
   c. Dilation
   d. Expulsion

_____ 25. Several genetic disorders use DNA fingerprinting in diagnosis, including
   a. Ovarian cancer
   b. Sickle cell anemia
   c. Lung cancer
   d. Squamous cell carcinoma

_____ 26. A long face, large ears, and flat feet are signs and symptoms of which disorder?
   a. Fragile X syndrome
   b. Down syndrome
   c. Albinism
   d. Multiple siblings

_____ 27. Pregnancy occurs when an ovum and a sperm cell unite. The prenatal period is divided into
   a. Four periods of two and a quarter months each
   b. Two periods of four and a half months each
   c. Three periods of three months each
   d. One nine-month period

_____ 28. By what week of gestation have the bones begun to harden?
   a. 5
   b. 12
   c. 30
   d. 28

_____ 29. When does the fetus start to gain substantial weight?
   a. Second month
   b. Fourth month
   c. Sixth month
   d. Eight month

_____ 30. After birth, which opening closes in most individuals and becomes the fossa ovalis?
   a. Foramen ovale
   b. Ductus arteriosus
   c. Ductus venosus
   d. Fertilization

_____ 31. The six weeks following birth is referred to as what period?
   a. Lactation period
   b. Dilation period
   c. Expulsion period
   d. Postnatal period

158   UNIT 2   Concepts of Common Illness by System

_____ 32. There is no cure for Turner's syndrome, a disorder that results when females
   a. Have a single X chromosome
   b. Have no ovaries
   c. Cannot synthesize enzymes
   d. Have Down syndrome

_____ 33. Which of the following are layers of cells formed in the inner cell mass of a blastocyst that form all of the embryo's organs?
   a. Empty cell layers
   b. Secondary germ layers
   c. Subcutaneous germ layers
   d. Primary germ layers

_____ 34. What act(s) as the primary structure for the exchange of substances between the fetus and mother?
   a. Enzymes
   b. Chorion
   c. Proteins
   d. Vitamins

_____ 35. Which structure allows nutrients and oxygen from maternal blood to pass to embryonic blood?
   a. Amnion
   b. Yolk sac
   c. Placenta
   d. Chorion

_____ 36. The cervix thins and softens in a process known as
   a. Expulsion
   b. Effacement
   c. Parturition
   d. Placental

_____ 37. The embryonic period of the prenatal period is identified as weeks
   a. 9 to delivery of the infant
   b. 2 through 8 of the pregnancy
   c. 0 to 4
   d. 13 to delivery

_____ 38. What is another term for Down syndrome, a disorder that causes mental retardation?
   a. PKU
   b. Amniocentesis
   c. Klinefelter's syndrome
   d. Trisomy 21

_____ 39. Inherited traits that are determined by multiple genes are called
   a. Complex inheritance
   b. Phenylalanine
   c. Implantation
   d. Growth hormones

_____ 40. The primary component of genes that is found in the nucleus of most cells is called
   a. Cytoplasm
   b. Achondroplastic dwarfism
   c. DNA
   d. Genetic anomalies

_____ 41. The uterus secretes which hormone to stimulate uterine contractions?
   a. Prostaglandins
   b. FSH
   c. Estrogen
   d. Progesterone

FILL IN THE BLANKS

In the space provided, write the word or phrase that best completes each sentence. Not all words or phrases are used.

42. The birth process ends _____.

43. The term for the first 28 days of the postnatal period is the _____.

44. After childbirth, _____ causes the mammary glands to produce milk.

45. _____ is the transfer of genetic traits from parent to child.

46. Albinism is a condition in which a person is born with _____ in the skin, eyes, or hair.

47. The _____ is called a zygote.

48. The fetal heart is _____ to the adult heart prior to birth.

49. The _____ chromosomal pair consists of one X chromosome and one Y chromosome in a male child.

50. After a baby is born, the ductus venosus normally closes to form the _____.

51. The placenta secretes large amounts of _____.

52. During pregnancy, the mother may experience calcium loss, _____, increased urination, and low back pain.

a. 22nd
b. 23rd
c. Fertilized egg
d. Human genetics
e. Identical
f. Increased cardiac output
g. Increased melanin
h. Ligamentum arteriosus
i. Ligamentum venosum
j. Little to no pigmentation
k. Neonatal period
l. Not identical
m. Oxytocin
n. Postnatal
o. Pregnancy
p. Progesterone and estrogen
q. Progesterone and prolactin
r. Prolactin

SHORT ANSWER

Write the answer to each statement or question on the lines provided.

53. Explain the three stages of the birth process, after the fetus has settled into position in the mother's pelvis.

_____

_____

_____

_____

54. What is the difference between genotype and phenotype?

55. What is DNA fingerprinting and how are some ways it currently is being used?

56. Identify the three periods of the prenatal period and the major developments that are associated with each period.

57. Describe the hormonal changes that occur in a woman during pregnancy.

LABELING

Follow the directions and write the answers on the lines provided.

58. Using the following illustration of the implantation of a blastocyst about six days after conception, identify these terms by writing them on the lines provided: *blastocyst, inner cell mass, trophoblast, uterine wall.*

   a. _____

   b. _____

   c. _____

   d. _____

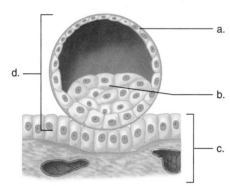

From *Hole's Human Anatomy & Physiology*, 12e, by Shier/Butler/and Lewis. Copyright © 2009. Reprinted by permission of McGraw-Hill Companies Inc.

59. In the following figure showing a developing fetus, identify these terms by writing them on the lines provided: *allantois, amnion, amniotic cavity, chorion, extraembryonic cavity, developing placenta, endometrium, maternal blood vessels, umbilical cord, yolk sac.*

   a. _____

   b. _____

   c. _____

   d. _____

   e. _____

   f. _____

   g. _____

   h. _____

   i. _____

   j. _____

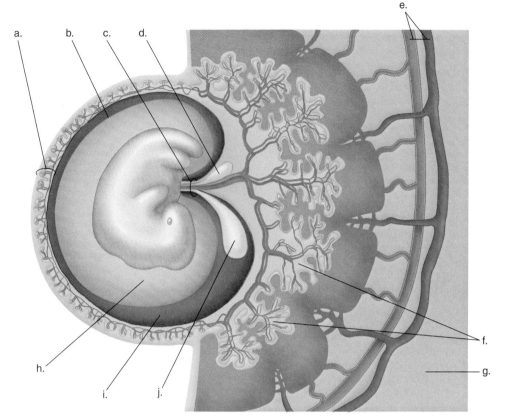

From *Hole's Human Anatomy & Physiology*, 12e, by Shier/Butler/and Lewis. Copyright © 2009. Reprinted by permission of McGraw-Hill Companies Inc.

critical thinking / application

CRITICAL THINKING

Write the answer to each statement on the lines provided.

60. Explain the difference between an embryo and a fetus.

APPLICATION

Write the answer to the question on the lines provided.

61. How may knowing your parent's genotype help with family planning?

case studies

Write your response to each case study statement or question on the lines provided.

62. A young couple is concerned with the possibility that their child whom the woman is currently pregnant with may have a genetic disorder. They would like you to explain the difference between choriocentesis and amniocentesis.

63. Why would exposure to measles during the first trimester of pregnancy be a greater risk to the fetus's nervous and cardiovascular systems than exposure in the third trimester?

pathophysiology

Follow the instructions for the statement.

64. Fill in the missing cells.

Disease	Etiology	Signs and Symptoms	Treatment
Cystic fibrosis			
Turner's syndrome			
Klinefelter's syndrome			
Down syndrome			

The Digestive System 19

vocabulary review

MATCHING

Match the key terms in the right column with the definitions in the left column by writing the letter of the correct answer in the space provided.

1. The projection off the posterior aspect of the soft palate that prevents food and liquids from entering the nasal cavity during swallowing
2. The throat
3. The opening between the pylorus of the stomach and the small intestine that controls movement of substances
4. The middle portion of the small intestine
5. The gastrointestinal tract and accessory organs
6. The double-walled outermost layer of the alimentary canal
7. The wormlike projection off the cecum that contains lymphoid tissue and has a role in immunity
8. Gums
9. "Baby" teeth
10. The act of chewing
11. Folds of the inner lining of the stomach that help mix the gastric contents
12. The varicosities of the rectum or anus
13. The first section of the small intestine that receives secretions from the pancreas that aid in digestion
14. The outermost layer of the lining of the GI tract
15. Small fingerlike projections that greatly increase the surface area of the small intestine so it can absorb many nutrients
16. The act of swallowing
17. The first portion of the large intestine
18. Broad, fanlike tissue that attaches to the posterior wall of the abdomen and holds the jejunum and ileum in the abdominal cavity
19. The last and longest section of the small intestine
20. A fold of mucous membrane that attaches two structures together

a. Alimentary canal
b. Cecum
c. Deciduous
d. Deglutition
e. Duodenum
f. Frenulum
g. Gingiva
h. Hemorrhoids
i. Ileum
j. Jejunum
k. Mastication
l. Mesentery
m. Microvilli
n. Parietal peritoneum
o. Pharynx
p. Pyloric sphincter
q. Rugae
r. Serosa
s. Uvula
t. Vermiform appendix

content review

MULTIPLE CHOICE

In the space provided, write the letter of the choice that best completes each statement or answers each question.

_____ 21. A fold of mucous membrane that attaches two structures is called the
   a. Serosa
   b. Gingiva
   c. Frenulum
   d. Mucosa

_____ 22. What is the valve that joins the last segment of the small intestine (ileum) with the first segment of the large intestine (cecum)?
   a. Ileocecal valve
   b. Appendiceal valve
   c. Jejunal valve
   d. Ileal valve

_____ 23. Which of the following is *not* a layer of the GI tract?
   a. Serosa
   b. Deciduous
   c. Adventitia
   d. Mucosa

_____ 24. The esophagus lies posterior to the
   a. Pharynx
   b. Trachea
   c. Larynx
   d. Stomach

_____ 25. The stomach functions to receive the food bolus from the
   a. Bronchi
   b. Trachea
   c. Esophagus
   d. Larynx

_____ 26. Which hormone made by the small intestine inhibits the gastric glands?
   a. Oxytocin
   b. PTH
   c. Gastrin
   d. CCK

_____ 27. The first section of the small intestine is called the
   a. Duodenum
   b. Jejunum
   c. Bowel
   d. Ileum

_____ 28. What joins the small and large intestine?
   a. Ileocecal valve
   b. Duodenal valve
   c. Colonic valve
   d. Splenic flexure

_____ 29. Which term means wormlike?
   a. Lentiform
   b. Canid
   c. Vermiform
   d. Cystic

_____ 30. The functions of the large intestine include
   a. Production of antibodies
   b. Production of certain vitamins
   c. Production of white blood cells
   d. Erythropoiesis

_____ 31. What is the approximate length of the rectum?
   a. 2 inches
   b. 8 inches
   c. 16 inches
   d. 20 inches

168   UNIT 2   Concepts of Common Illness by System

_____ **32.** Feces are made of undigested solid materials, a little water, ions, mucus, cells of the intestinal lining, and
   a. Antibodies
   b. Red blood cells
   c. White blood cells
   d. Bacteria

_____ **33.** The back of the tongue contains lymphatic tissue called
   a. Lingual tonsils
   b. Vermiform appendix
   c. Peyer's patches
   d. Adenoids

_____ **34.** Where are the soft and hard palates found?
   a. In the floor of the mouth
   b. In the walls of the nasal cavity
   c. In the roof of the mouth
   d. In the roof of the nasal cavity

_____ **35.** Salivary glands secrete saliva, which is a mixture of water, enzymes, and which of the following?
   a. Hormones
   b. Mucus
   c. Nucleic acids
   d. Pepsin

_____ **36.** Which glands are located in the floor of the mouth?
   a. Palatine
   b. Submandibular
   c. Parotid
   d. Endocrine

_____ **37.** What develops when an organ pushes through a wall that contains it?
   a. Hernia
   b. Aneurysm
   c. Subluxation
   d. Dislocation

_____ **38.** What mutated gene causes cystic fibrosis?
   a. CFTR
   b. Brc
   c. Apoptotic
   d. Sickle cell

_____ **39.** Which organ lies below the diaphragm in the left upper quadrant of the abdominal cavity?
   a. Heart
   b. Liver
   c. Gallbladder
   d. Stomach

_____ **40.** What separates the oral cavity from the nasal cavity?
   a. Hard palate
   b. Diaphragm
   c. Nasal conchae
   d. Vomer and ethmoid bones

_____ **41.** The middle portion of the small intestine is also known as the
   a. Bowel
   b. Duodenum
   c. Ileum
   d. Jejunum

FILL IN THE BLANKS

In the space provided, write the word or phrase that best completes each sentence. Not all words or phrases are used.

42. The _____, not the stomach, absorbs many substances.
43. Gastritis is a(n) _____ of the stomach lining.
44. Smooth muscle in the wall of the _____ produces churning and peristalsis.
45. Bile leaves the liver through the _____.
46. The cecum is the first portion of the _____.
47. The appendix is located in the _____ of the abdominal cavity.
48. The last inch of the rectum is known as the _____.
49. The _____ fits in the "C" of the duodenum.
50. Pancreatic amylase digests _____.
51. The _____ ligament of the liver is a fold of mesentery.
52. The liver produces many _____ including albumin and fibrinogen.

a. Anus
b. Carbohydrates
c. Common bile duct
d. Falciform
e. Hepatic
f. Hepatic duct
g. Inflammation
h. Large intestine
i. Left lower quadrant
j. Lipids
k. Necrosis
l. Pancreas
m. Proteins
n. Right upper quadrant
o. Right lower quadrant
p. Small intestine

SHORT ANSWER

Write the answer to each statement or question on the lines provided.

53. Identify what vitamins the liver stores.

_____

_____

_____

_____

170   UNIT 2   Concepts of Common Illness by System

54. Describe the effects of aging on the digestive system.

55. Explain the causes, signs and symptoms, and treatments of colon cancer.

56. Explain how the rectum differs from the anal canal.

57. What does it mean for a person to be tongue-tied?

LABELING

Follow the directions and write the answers on the lines provided.

58. Using the following illustration showing the organs of the digestive system, identify these terms by writing them on the lines provided: *anus, esophagus, gallbladder, large intestine, liver, mouth, pancreas, pharynx, rectum, salivary glands, small intestine, stomach.*

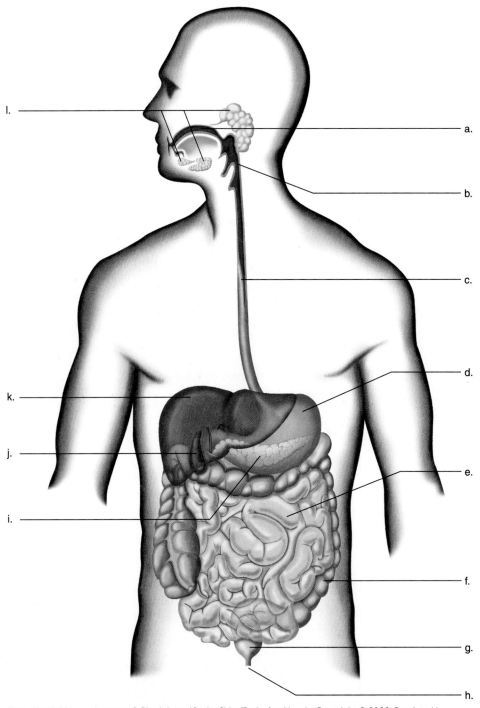

From *Hole's Human Anatomy & Physiology*, 12e, by Shier/Butler/and Lewis. Copyright © 2009. Reprinted by permission of McGraw-Hill Companies Inc.

a. _____

b. _____

c. _____

d. _____

e. _____

f. _____

g. _____

h. _____

i. _____

j. _____

k. _____

l. _____

59. Using the following figure showing the different regions of the stomach, identify these terms by writing them on the lines provided: *body, cardia, duodenum, esophagus, fundus, pyloric antrum, pyloric canal, pyloric sphincter, pylorus, rugae.*

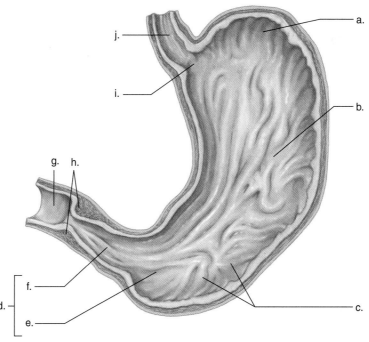

From *Hole's Human Anatomy & Physiology*, 12e, by Shier/Butler/and Lewis. Copyright © 2009. Reprinted by permission of McGraw-Hill Companies Inc.

a. _____

b. _____

c. _____

d. _____

e. _____

f. _____

g. _____

h. _____

i. _____

j. _____

60. Using the following illustration showing the various parts of the large intestine and the layers in the wall of the alimentary canal, identify these terms by writing them on the lines provided: *anal canal, appendix, ascending colon, cecum, descending colon, haustra, ileocecal sphincter, ileum, muscular layer, mucous membrane, orifice of appendix, rectum, serous layer, sigmoid colon, transverse colon.*

a. _____

b. _____

c. _____

d. _____

e. _____

f. _____

g. _____

h. _____

i. _____

j. _____

k. _____

l. _____

m. _____

n. _____

o. _____

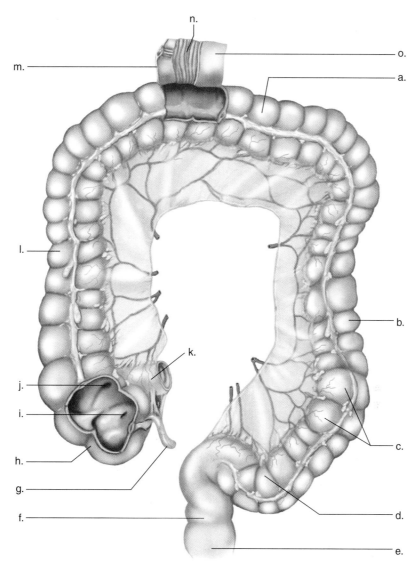

From *Hole's Human Anatomy & Physiology*, 12e, by Shier/Butler/and Lewis. Copyright © 2009. Reprinted by permission of McGraw-Hill Companies Inc.

critical **thinking** / **application**

CRITICAL THINKING

Write the answer to each statement on the lines provided.

61. Describe the structure and function of the pancreas.

_____
_____
_____
_____

62. Explain how the esophagus propels food.

63. Explain the process of swallowing.

APPLICATION

Write the answer to the statement on the lines provided.

64. Explain why someone who has had his or her gallbladder removed must eat smaller amounts of fatty foods.

case studies

Write your response to each case study question or statement on the lines provided.

65. A 3-year-old boy is having difficulty with his speech because he is tongue-tied. What is the medical term for this and what can be done to help the boy?

66. A 62-year-old man has never had a colonoscopy. He knows that a grandfather and uncle both had colon cancer. Explain the importance of having this procedure done.

_____

_____

_____

_____

pathophysiology

Follow the instructions for the statement.

67. Fill in the missing cells.

Disease	Etiology	Signs and Symptoms	Treatment
Oral cancer			
Gastritis			
Crohn's disease			

Metabolic Function and Nutrition

vocabulary review

MATCHING

Match the key terms in the right column with the definitions in the left column by writing the letter of the correct answer in the space provided.

_____ 1. A molecule formed from one molecule of glycerol and three fatty acid molecules

_____ 2. A metabolic reaction that involves the breaking down of larger molecules into smaller molecules

_____ 3. A category of food that includes starches, simple sugars, and cellulose

_____ 4. An anaerobic process of eight reactions that occurs in the mitochondria and produces CO_2, ATP, NADH, and $FADH_2$

_____ 5. The conversion of glycogen stored in the liver and skeletal muscle into glucose

_____ 6. The amount of energy used by the body when it is in the resting state

_____ 7. A type of starvation that entails both protein and calorie insufficiency

_____ 8. A process in which a pyruvic acid molecule loses a CO_2 molecule

_____ 9. Four sets of reactions that involve oxidation of glucose

_____ 10. A metabolic reaction that involves the synthesis of substances

_____ 11. A molecule formed in the matrix of the mitochondria that is the substrate that enters the Krebs cycle

_____ 12. An aerobic series of electron transfers that occurs on the inner membrane of the mitochondria resulting in the production of 32 ATP molecules for every glucose molecule that is oxidized

_____ 13. A type of starvation in which there is too little protein in the diet

_____ 14. The formation of glucose from proteins and fats

_____ 15. Fats

a. Acetyl coenzyme A
b. Anabolic reactions
c. Basal metabolic rate
d. Carbohydrates
e. Catabolic reactions
f. Cellular respiration
g. Conduction
h. Decarboxylation
i. Electron transport chain
j. Gluconeogenesis
k. Glycogenolysis
l. Krebs cycle
m. Kwashiorkor
n. Lipids
o. Marasmus
p. Radiation
q. Triglycerides

content review

MULTIPLE CHOICE

In the space provided, write the letter of the choice that best completes each statement or answers each question.

_____ 16. Which of the following is a lipoprotein that is made in the liver?
   a. Glycerol
   b. ATP
   c. LDL
   d. Anabolic

_____ 17. Which type of metabolic reaction involves the decomposition or breakdown of complex molecules into simpler molecules?
   a. Cholesterol
   b. Catabolic
   c. Fiber diets
   d. Adenomas

_____ 18. ATP is the _____ of the cell.
   a. Cholesterol
   b. Genetic information
   c. Protein
   d. Energy currency

_____ 19. Which of the following is also known as the citric acid cycle or tricarboxylic cycle?
   a. Krebs cycle
   b. Adipose tissue
   c. Carbohydrate metabolism
   d. Glycogen storage

_____ 20. Skeletal muscle can store up to three times as much glycogen as the
   a. Lungs
   b. Heart
   c. Liver
   d. Kidney

_____ 21. The combination of lipids and proteins is called
   a. Lipoprotein
   b. Catabolic reactions
   c. ATP
   d. Homeostasis

_____ 22. Which type of metabolic reaction involves synthesis of substances?
   a. Catabolic
   b. Anabolic
   c. Adenosine diphosphate
   d. Carbohydrate

_____ 23. In an adult, LDL cholesterol should be less than
   a. 80 to 100 mg/dL
   b. 130 mg/dL
   c. 4 ATP molecules
   d. 32 ATP molecules

_____ 24. Which of the following is (are) the main energy currency of the cell?
   a. Electron transport chain
   b. Lipids
   c. ATP
   d. Lipoproteins

_____ 25. Protein is used by the body for
   a. Growth and the repair of tissues
   b. Transport glycogenolysis
   c. Anaerobic reaction
   d. Cellular respiration

_____ 26. Carbohydrates include cellulose, starches, and
   a. Liver enzymes
   b. Proteins
   c. Fats
   d. Simple sugars

_____ 27. Triglycerides are *not* found in
   a. Water
   b. Eggs

c. Milk
d. Butter

____ 28. When glycogen storage capacity is met, what can excess glucose be converted to?
   a. Glycerol and fatty acids
   b. Balanced metabolism
   c. Phosphates
   d. Homeostasis

____ 29. What energy-producing process occurs in the cytoplasm of the cell and does not require oxygen and so is called an anaerobic reaction?
   a. Homeostasis
   b. Metabolism
   c. Glycolysis
   d. ATP

____ 30. One gram of carbohydrate equals
   a. 1 calorie of energy
   b. 16 calories of energy
   c. 4 calories of energy
   d. 1 protein

____ 31. Lipids are necessary for cell membranes, many hormones, blood clotting, and
   a. Arteriosclerosis
   b. Lipid-lowering agents
   c. Development of adipose tissue
   d. Myelin that surrounds many nerves

____ 32. Which of the following is true about hypercholesterolemia?
   a. It decreases the risk of heart disease
   b. It results from excess cholesterol in the blood
   c. It is viral in nature
   d. It is the equivalent of 9 calories of energy

____ 33. A patient displays hypertension, hyperinsulinism, excess body fat, and/or hypercholesterolemia. Which disorder does he or she have?
   a. Turner's syndrome
   b. Metabolic syndrome
   c. Homeostasis syndrome
   d. ATP

____ 34. Heat is lost from the body through all of the following *except*
   a. Shivering
   b. Convection
   c. Evaporation
   d. Radiation

____ 35. Homeostasis means the right amount of nutrients are
   a. Raising the body temperature
   b. Changed into imbalance
   c. Obtained and used
   d. Sending shutdown signals to the hypothalamus

____ 36. What is lost through evaporation, conduction, radiation, and convection?
   a. Oocytes
   b. Water
   c. ATP
   d. Heat

____ 37. To have a balanced diet we need
   a. Lipid, protein, and carbohydrate nutrients
   b. ATP
   c. Water and oil
   d. Vitamins C and D

____ 38. Vitamin D is essential for the absorption of
   a. Lipids
   b. Water
   c. Calcium and phosphorus
   d. Proteins

CHAPTER 20   Metabolic Function and Nutrition

_____ **39.** Vitamin A is essential for epithelial cells, acting as a(n)
   a. Riboflavin
   b. Antioxidant
   c. Pellagra
   d. ATP

_____ **40.** Vitamin B_6 is an important
   a. Coenzyme
   b. Metabolic syndrome
   c. Cyanocobalamin
   d. ATP

_____ **41.** Which of the following best describes minerals?
   a. They are carbohydrates
   b. They are lipids
   c. They are inorganic substances
   d. They are organic substances

FILL IN THE BLANKS

In the space provided, write the word or phrase that best completes each sentence. Not all words and phrases are used.

42. Metabolism is the sum of all _____ in the body.

43. _____ involve the breaking down of large molecules into smaller molecules.

44. Anabolic and catabolic reactions must be in _____ for the body to be in homeostasis.

45. _____ provides fiber or bulk for the large intestine which helps the large intestine empty regularly.

46. Normal blood glucose level is _____ of plasma.

47. Energy is not required for the entry of _____ into the cells.

48. The Krebs cycle occurs in mitochondria and since oxygen is required, it is classified as a(n) _____ process.

49. _____ have the highest percentage of proteins.

50. HDLs transport cholesterol to the _____ for elimination and are called "good" cholesterol.

51. There are _____ amino acids in the human body.

52. Carbohydrates, fats, and _____ are the three major food categories.

a. 100 to 120 mg/100 mL
b. 20
c. 60 to 80 mg/100 mL
d. 80 to 100 mg/100 mL
e. Aerobic
f. Anaerobic
g. Balance
h. Catabolic reaction
i. Cellulose
j. Chemical reactions
k. Glucose
l. High-density lipoproteins
m. Liver
n. Low-density lipoproteins
o. Pancreas
p. Proteins
q. Starch
r. Stomach
s. Sucrose

SHORT ANSWER

Write the answer to each statement or question on the lines provided.

53. Explain the differences between anabolic and catabolic reactions.

54. Explain the Krebs cycle.

55. What are the substances that make up LDLs and VLDLs?

56. Identify the normal cholesterol levels that occur in an adult.

CHAPTER 20 Metabolic Function and Nutrition

LABELING

Follow the directions and write the answers on the lines provided.

57. Using the following figure showing a molecule of ATP, identify these terms by writing them on the lines provided: *adenosine, adenine, phosphates, ribose*.

a. _____

b. _____

c. _____

d. _____

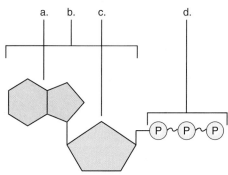

From *Hole's Human Anatomy & Physiology*, 12e, by Shier/Butler/and Lewis. Copyright © 2009. Reprinted by permission of McGraw-Hill Companies Inc.

58. The electron transport chain takes place on the inner membrane of mitochondria, as illustrated in the following figure. Identify these terms by writing them on the lines provided: *cristae, inner membrane, intermembrane space, matrix, outer membrane*.

a. _____

b. _____

c. _____

d. _____

e. _____

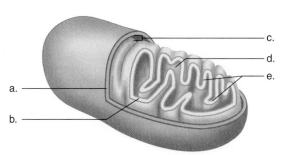

From *Hole's Human Anatomy & Physiology*, 12e, by Shier/Butler/ and Lewis. Copyright © 2009. Reprinted by permission of McGraw-Hill Companies Inc.

critical thinking / application

CRITICAL THINKING

Write the answer to each statement on the lines provided.

59. Explain the electron transport chain.

APPLICATION

Write the answer to the statement on the lines provided.

60. Explain the concept of basal metabolic rate (BMR) and its connection to weight gain or weight loss.

case studies

Write your response to each case study statement on the lines provided.

61. A 58-year-old man with a history of heart disease must lower his lipid levels. Explain the different types of lipids and appropriate serum levels.

62. A young athlete is on an all-protein diet. Explain why an individual needs carbohydrates and fats in his or her diet as well as proteins.

pathophysiology

Follow the instructions for the statement.

63. Fill in the missing cells.

Disease	Etiology	Signs and Symptoms	Treatment
Hypercholesterolemia			
Pellagra			
Anorexia nervosa			
Osteomalacia			

The Endocrine System

21

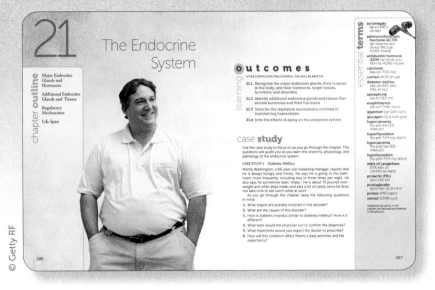

vocabulary review

MATCHING

Match the key terms in the right column with the definitions in the left column by writing the letter of the correct answer in the space provided.

_____ 1. A lipid-soluble hormone built of cholesterol
_____ 2. A protein hormone that moves calcium from the blood to bone
_____ 3. A lipid-soluble hormone that has a role in inflammation and intensifying pain
_____ 4. An excess secretion of thyroid hormone
_____ 5. An abnormal increase in the length of long bones due to hypersecretion of growth hormone during childhood
_____ 6. A condition caused by too much growth hormone secreted after puberty
_____ 7. The peptide hormone responsible for milk production in the mammary glands
_____ 8. A peptide hormone that causes the body to retain water by increasing water resorption by the kidneys
_____ 9. A condition in which the eyeballs protrude, often caused by hyperthyroidism
_____ 10. A condition in which there is too little calcium in the blood
_____ 11. An amino acid hormone secreted by the adrenal medulla
_____ 12. A glucocorticoid released by the adrenal cortex in response to stress
_____ 13. A hormone secreted by the pancreas that increases blood glucose levels
_____ 14. Groups of cells in the pancreas that have an endocrine function
_____ 15. A peptide hormone secreted by the pancreas that lowers blood glucose levels

a. Acromegaly
b. Antidiuretic hormone
c. Calcitonin
d. Cortisol
e. Diabetes mellitus
f. Epinephrine
g. Exophthalmos
h. Gigantism
i. Glucagon
j. Hypercalcemia
k. Hyperthyroidism
l. Hypocalcemia
m. Insulin
n. Islets of Langerhans
o. Prolactin
p. Prostaglandin
q. Steroid

content review

MULTIPLE CHOICE

In the space provided, write the letter of the choice that best completes each statement or answers each question.

_____ 16. Which of the following is a bony structure that protects the pituitary gland?
   a. The optic chiasm
   b. The sella turcica
   c. The paranasal sinus
   d. The larynx

_____ 17. Which gland is also known as the adenohypophysis?
   a. The pineal body
   b. The posterior pituitary
   c. The hypothalamus
   d. The anterior pituitary

_____ 18. Which of the following are small glands that are embedded within another gland?
   a. The adrenal glands
   b. The ovaries
   c. The parathyroid glands
   d. The pituitary glands

_____ 19. Which of the following is a hormone produced by the pineal gland, regulates the biological clock, and is linked to the onset of puberty?
   a. Gastrin
   b. Growth hormone
   c. Melatonin
   d. ADH

_____ 20. Which of the following is also known as hypocortisolism?
   a. Cushing's syndrome
   b. Graves' disease
   c. Acromegaly
   d. Addison's disease

_____ 21. Which cells of the islets of Langerhans secrete insulin?
   a. Alpha
   b. Beta
   c. Gastric
   d. Cortex

_____ 22. Which type of hormones are built from cholesterol?
   a. Steroidal
   b. Nonsteroidal
   c. Lipid
   d. Prostaglandins

_____ 23. Hormones from which of the following glands are responsible for T lymphocyte maturation?
   a. The thymus
   b. The thyroid
   c. The pancreas
   d. The parathyroid

_____ 24. People who are on long-term steroid therapy are at increased risk for which of the following diseases?
   a. Diabetes mellitus
   b. Diabetes insipidus
   c. Hypothyroidism
   d. Renal failure

_____ 25. Which of the following diseases may be congenital in nature?
   a. Myxedema
   b. Graves' disease
   c. Cretinism
   d. Addison's disease

_____ 26. Which of the following is an example of a positive feedback?
   a. Labor and delivery
   b. Insulin release when glucose levels in the blood are high
   c. Glucagon release when glucose levels in the blood are low
   d. Calcitonin release when blood calcium levels are high

_____ 27. The hypothalamus produces
   a. Oxytocin and ADH
   b. Growth hormone
   c. Luteinizing hormone and follicle-stimulating hormone
   d. Melatonin

_____ 28. TSH stimulates
   a. The thyroid to release thyroid hormone
   b. The testes to release their hormones
   c. The thalamus to release melatonin
   d. The pineal gland to release melatonin

_____ 29. Parathyroid hormone helps tear down bone by
   a. Stimulating osteoclasts
   b. Stimulating osteoblasts
   c. Stimulating osteocytes
   d. Stimulating growth hormone

_____ 30. Which of the following does the adrenal cortex secrete?
   a. Cortisol
   b. Epinephrine
   c. Melatonin
   d. Insulin

_____ 31. The uptake of glucose by cells is promoted by
   a. Insulin
   b. Glucagon
   c. Melatonin
   d. Luteinizing hormone

_____ 32. Which disorder is produced in children by too little thyroid hormone being secreted?
   a. Cretinism
   b. Acromegaly
   c. Giantism
   d. Diabetes mellitus

_____ 33. Hyposecretion of ADH produces
   a. Diabetes insipidus
   b. Diabetes mellitus
   c. Acromegaly
   d. Addison's disease

_____ 34. Hyposecretion of cortisol produces
   a. Addison's disease
   b. Exophthalmos
   c. Cushing's syndrome
   d. Acromegaly

_____ 35. The combination of rising PTH with declining calcitonin leads to increased
   a. Osteoporosis
   b. Acromegaly
   c. Diabetes insipidus
   d. Diabetes mellitus

FILL IN THE BLANKS

In the space provided, write the word or phrase that best completes each sentence. Not all words and phrases are used.

36. The target cells of a hormone are the cells that contain the _____ for the hormone.

37. A stimulus that produces a physiological change in the body is a(n) _____.

38. Glucagon is produced by the _____ cells of the pancreas.

39. Release of _____ by the posterior pituitary increases uterine contractions during delivery.

40. Hormone levels are controlled by _____.

41. Cretinism is caused by a deficiency of _____.

42. Hypersecretion of ACTH results in _____.

43. A decrease in ADH secretion will result in _____.

44. Luteinizing hormone is secreted by the _____.

45. The _____ is located in the mediastinum.

a. Addison's disease
b. Alpha
c. Anterior pituitary gland
d. Antidiuretic hormone
e. Beta
f. Cushing's syndrome
g. Diabetes insipidus
h. Effector
i. Feedback loops
j. Oxytocin
k. Posterior pituitary
l. Prolactin
m. Receptor
n. Spleen
o. Stressor
p. Thymus
q. Thyroid hormone

SHORT ANSWER

Follow the directions and write the answer to each statement on the lines provided.

46. Describe the following disorders and the signs and symptoms of each.

   a. Acromegaly _____

   _____

   b. Cushing's syndrome _____

   _____

   c. Graves' disease _____

   _____

   d. Myxedema _____

   _____

   e. Diabetes mellitus _____

   _____

47. Answer the following questions about diabetes mellitus.

   a. What is the difference between type 1 and type 2 diabetes? _____

   _____

   b. What are the causes of diabetes? _____

   _____

   c. What are some treatments for diabetes? _____

   _____

LABELING

Follow the directions and write the answers on the lines provided.

48. Using the following figure illustrating the major endocrine glands, identify these terms by writing them on the lines provided: *adrenal gland, hypothalamus, kidney, ovary, pancreas, parathyroid gland, pineal gland, pituitary gland, testes, thymus, thyroid gland.*

a. _____

b. _____

c. _____

d. _____

e. _____

f. _____

g. _____

h. _____

i. _____

j. _____

k. _____

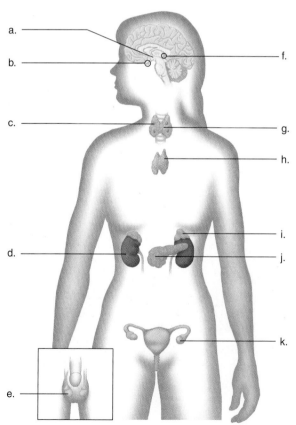

From *Hole's Human Anatomy & Physiology*, 12e, by Shier/Butler/and Lewis. Copyright © 2009. Reprinted by permission of McGraw-Hill Companies Inc.

49. Using the following illustration showing the pituitary gland and associated structures, identify these terms by writing them on the lines provided: *anterior lobe, hypothalamus, pituitary stalk (infundibulum), posterior lobe, sella turcica, sphenoid bone, sphenoidal sinus.*

a. _____

b. _____

c. _____

d. _____

e. _____

f. _____

g. _____

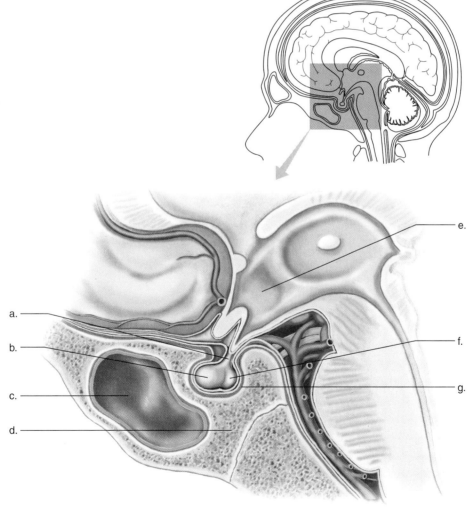

From *Hole's Human Anatomy & Physiology*, 12e, by Shier/Butler/and Lewis. Copyright © 2009. Reprinted by permission of McGraw-Hill Companies Inc.

CHAPTER 21   The Endocrine System

50. Using the following figure that illustrates the thyroid gland and associated structures, identify these terms by writing them on the lines provided: *colloid, extrafollicular cell, follicular cell, isthmus, larynx, thyroid gland.*

   a. _____

   b. _____

   c. _____

   d. _____

   e. _____

   f. _____

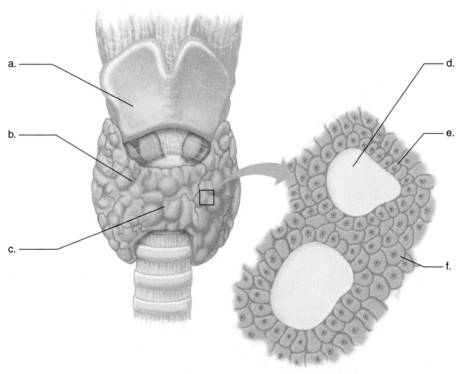

From *Hole's Human Anatomy & Physiology,* 12e, by Shier/Butler/and Lewis. Copyright © 2009. Reprinted by permission of McGraw-Hill Companies Inc.

51. Using the following illustration showing the pancreas and associated structures, identify these terms by writing them on the lines provided: *common bile duct, gallbladder, pancreas, pancreatic duct,* and *small intestine.*

a. _____

b. _____

c. _____

d. _____

e. _____

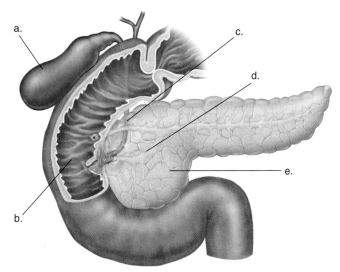

From *Hole's Human Anatomy & Physiology,* 12e, by Shier/Butler/and Lewis. Copyright © 2009. Reprinted by permission of McGraw-Hill Companies Inc.

critical thinking / application

CRITICAL THINKING

Write the answer to each question or statement on the lines provided.

52. Why must nonsteroidal hormones have their receptors on cell surfaces and not inside cells?

53. Why would a person have thyroid problems if his or her anterior pituitary gland was damaged?

54. Given the information you know about cortisol and steroidal hormones, explain why patients taking steroids such as prednisone complain of weight gain, tiredness, and extreme hunger.

55. How do calcitonin and parathyroid hormone work together to control calcium levels in the blood?

56. Why is the pancreas considered both an endocrine and an exocrine gland?

_____

_____

_____

_____

APPLICATION

Follow the directions and write the answer to each statement on the lines provided.

57. Hormone Targets: Hormones do not affect all cells in the body, only their target cells. For each hormone listed below, give its target(s).

  a. ACTH _____

  b. Insulin _____

  c. TSH _____

  d. Calcitonin _____

  e. FSH and LH _____

  f. MSH _____

58. Endocrine Disorders: Endocrine disorders can produce too many or too few hormones. Name the disorders that may result from the following conditions.

  a. Hypersecretion of GH in adults _____

  b. Hyposecretion of GH in children _____

  c. Hypersecretion of thyroid hormone in adults _____

  d. Hyposecretion of thyroid hormone in adults _____

  e. Hyposecretion of thyroid hormone in children _____

  f. Hyposecretion of ACTH (adults or children) _____

case studies

Write your response to each case study question on the lines provided.

59. A female patient has a diet that is totally lacking in lipids. How does her diet affect hormone production in her body?

_____

_____

_____

_____

60. A 35-year-old woman is pregnant with her first child. Her glucose tolerance test (GTT) is elevated and she has been put on a diabetic diet. She is relieved that her diabetes is gestational and that she will return to normal blood sugar levels after her pregnancy. What information might you want to share with her?

_____

_____

_____

_____

pathophysiology

Follow the instructions for the statement.

61. Fill in the missing cells.

Disease	Etiology	Signs and Symptoms	Treatment
Graves' disease			
Hyperparathyroidism			
Diabetes mellitus			

The Special Senses 22

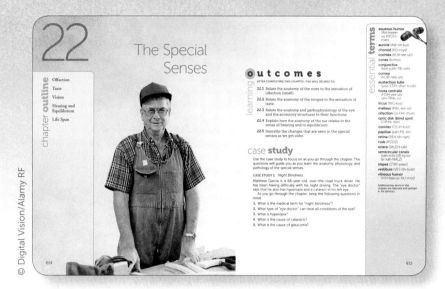

vocabulary review

MATCHING

Match the key terms in the right column with the definitions in the left column by writing the letter of the correct answer in the space provided.

1. The stirrup-shaped bone of the middle ear
2. The inner layer of the eye that senses light
3. The spiral structure of the ear that contains hearing receptors
4. Photoreceptors that detect blue, green, and red and provide sharper images than rods
5. The bones of the middle ear, the smallest bones in the body
6. Responsible for night and peripheral (side) vision
7. The area of the retina where the optic nerve enters and where there are no sensory nerves; also known as the blind spot
8. A pathway between the throat and the middle ear that allows organisms to pass between them
9. A watery fluid that fills the eye's anterior and posterior chambers
10. This surface location is where taste buds are found
11. A thick, jellylike substance
12. It forms the "white" of the eye
13. Bends light as it enters the eye
14. The anvil-shaped bone of the middle ear
15. The sense of smell

a. Aqueous humor
b. Auricle
c. Choroid
d. Cochlea
e. Cones
f. Conjunctiva
g. Cornea
h. Eustachian tube
i. Fovea centralis
j. Incus
k. Malleus
l. Olfaction
m. Optic disk
n. Ossicles
o. Papillae
p. Retina
q. Rods
r. Sclera
s. Semicircular canals
t. Stapes
u. Vestibule
v. Vitreous humor

199

content review

MULTIPLE CHOICE

In the space provided, write the letter of the choice that best completes each statement or answers each question.

_____ 16. Which of the following is *not* a special sense organ?
   a. Nose
   b. Eyes
   c. Tongue
   d. Lips

_____ 17. Taste cells that respond to sour chemicals are concentrated on what part of the tongue?
   a. The tip
   b. The back
   c. The sides
   d. The bottom

_____ 18. Where is the organ of Corti located?
   a. In the cochlea
   b. In the vestibule
   c. In the tympanic membrane
   d. In the semicircular canals

_____ 19. What is the term for a physician who specializes in treating diseases and conditions of the eye?
   a. Optician
   b. Otologist
   c. Optometrist
   d. Ophthalmologist

_____ 20. Which of the following indicates abnormal near vision?
   a. Hyperopia
   b. Myopia
   c. Amblyopia
   d. Astigmatism

_____ 21. The auricle begins the hearing process by collecting sound waves and channeling them to
   a. The labyrinth
   b. The tympanic membrane
   c. The cochlea
   d. The pinna

_____ 22. Why do physicians use tuning forks?
   a. To measure air pressure in the ear
   b. To treat hearing loss
   c. To determine the extent of hearing loss
   d. To determine if the patient has hearing loss

_____ 23. Which condition is caused by an accumulation of aqueous humor in the eye?
   a. Glaucoma
   b. Conjunctivitis
   c. Otitis media
   d. Tympanic membrane rupture

_____ 24. The most common cause for vision loss in the United States is
   a. Cataracts
   b. Glaucoma
   c. Retinitis
   d. Macular degeneration

_____ 25. Which disorder is caused by an abnormally shaped cornea or lens?
   a. Astigmatism
   b. Amblyopia
   c. Strabismus
   d. Entropion

FILL IN THE BLANKS

In the space provided, write the word or phrase that best completes each sentence. Not all words or phrases are used.

26. When you eat spicy foods, you are activating _____ receptors on the tongue.

27. The fifth acknowledged taste sensation is _____.

28. _____ is the medical term for crossed eyes and walleyes.

29. The _____ controls the amount of light entering the eye.

30. Loss of visual acuity due to aging is called _____.

31. A(n) _____ is used to measure hearing acuity.

32. The _____ send information about the position of the head along the vestibular nerves to the cerebrum for interpretation.

33. _____ is an eversion of the lower eyelid.

34. The test that is done to measure the eardrum's ability to move and thus gauges pressure in the middle ear is called _____.

35. The _____ is the visual professional who fills eyeglass and contact lens prescriptions.

36. The middle layer of the eyeball is the _____.

37. The corneal–scleral junction is also called the _____.

38. _____ control the shape of the lens, making the lens more or less curved for viewing either near or distant objects, respectively.

39. Tears are produced by the _____.

40. The taste cells that respond to _____ chemicals or tastes are located on the tip of the tongue.

a. Audiometer
b. Bitter
c. Cornea
d. Ectropion
e. Equilibrium receptors
f. Lacrimal gland
g. Limbus
h. Muscles of the ciliary body
i. Ophthalmologist
j. Optician
k. Pain
l. Presbyopia
m. Pupil
n. Strabismus
o. Sweat gland
p. Sweet
q. Touching
r. Tympanometry
s. Umami
t. Uvea

SHORT ANSWER

Follow the directions and write the answers on the lines provided.

41. Describe each of the following conditions of the special senses in the lines provided.

   a. Glaucoma _____

   _____

   _____

   b. Entropion _____

   _____

   _____

   c. Dry eye syndrome _____

   _____

   _____

   d. Retinal detachment _____

   _____

   _____

LABELING

Follow the directions and write the answers on the lines provided.

42. Using the following figure showing a horizontal section of the eye, identify these terms by writing them on the lines provided: *anterior chamber, choroid, CNII (optic), cornea, iris, lens, optic disc (blind spot), pupil, retina, sclera, vitreous chamber (posterior cavity)*.

   a. _____

   b. _____

   c. _____

d. _____
e. _____
f. _____
g. _____
h. _____
i. _____
j. _____
k. _____

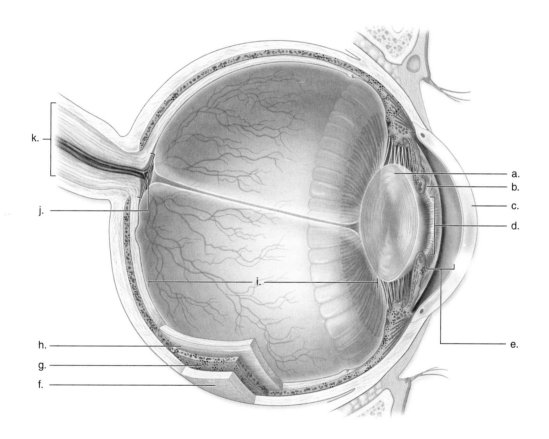

CHAPTER 22　The Special Senses　203

43. Using the following figure showing the outer, middle, and inner ear, identify these terms by writing them on the lines provided: *auricle, semicircular canals, incus, malleus, stapes, cochlea, vestibulocochlear nerve, oval window (under stapes), round window, tympanic cavity, tympanic membrane, external acoustic meatus, auditory tube, pharynx.*

a. _____

b. _____

c. _____

d. _____

e. _____

f. _____

g. _____

h. _____

i. _____

j. _____

k. _____

l. _____

m. _____

n. _____

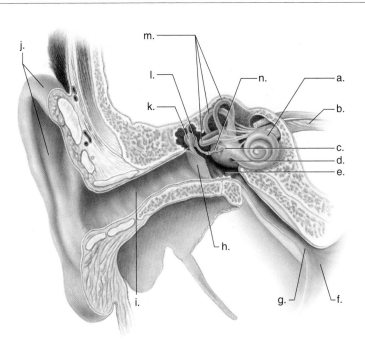

From *Hole's Human Anatomy & Physiology*, 12e, by Shier/Butler/and Lewis. Copyright © 2009. Reprinted by permission of McGraw-Hill Companies Inc.

critical **thinking** / **application**

CRITICAL THINKING

Write the answer to each question on the lines provided.

44. If a person is on a drug that causes his or her nasal membranes to become dry, why might he or she have trouble smelling?

45. How might a sore throat produce otitis media?

46. Why is it important to face a person with a hearing impairment instead of speaking directly into his or her ear?

47. How might the speech development of an infant with chronic otitis media be affected?

APPLICATION

Follow the directions and write the answer to each statement on the lines provided.

48. Eye Safety: The overwhelming majority of eye injuries can be avoided by using eye safety practices. Name some eye safety practices for each situation.

    a. At home _____

    b. While playing sports _____

    c. In the workplace _____

case studies

Write your response to each case study question on the lines provided.

49. A 65-year-old woman has been diagnosed with glaucoma. The doctor prescribes eye drops to treat the condition. However, the woman complains that she does not like putting medicines in her eyes and that her eyes do not hurt. How would you convince her that she needs to use the eye drops?

50. A 52-year-old woman has noticed that when she looks at straight lines, they sometimes become wavy. She also has noticed that she cannot see details very well and that her central vision is becoming worse. What condition does she most likely have?

_____

_____

_____

_____

pathophysiology

Follow the instructions for the statement.

51. Fill in the missing cells.

Disease	Etiology	Signs and Symptoms	Treatment
Amblyopia			
Macular degeneration			
Ectropion			
Meniere's disease			